A Field Guide to Community Based Adaptation

The world's poor will be the most critically affected by a changing climate—and yet their current plight is not improving rapidly enough to fulfill the UN's Millennium Development Goals. If experienced development organizations are finding it difficult to solve decades-old development problems, how will they additionally solve new challenges driven by climate change? *A Field Guide to Community Based Adaptation* illustrates how including community members in project design and co-management leads to long-lasting, successful achievement of development and adaptation goals.

This field guide provides a system of building-block activities for staff on the ground to use in developing and implementing successful adaptation to climate change projects that can be co-managed and sustained by communities. Based on years of use in 129 different countries, the techniques illustrated in this field guide use a step-by-step progression to lead readers through problem assessment, project design, implementation, and community take-over. The book equips development staff with all the tools and techniques they need to improve current project effectiveness, to introduce community based adaptation into organizational programming and to generate new projects. The techniques provided can be applied to a broad range of challenges, from agriculture and soil and water challenges, to health concerns, flood defenses, and market development. The book is supported by a user-friendly website (http://www.timmagee.net/field-guide-to-cba/) updated by the author, where readers can download online resources for each chapter which they can tailor to their own specific projects.

This practical guide is accessible to all levels of development staff and practitioners, as well as to students of development and environmental studies.

Tim Magee is the Executive Director of the Center for Sustainable Development, Inc., and is based in Guatemala. He has provided online instruction on community-centered, sustainable development to practitioners from 380 organizations in 129 countries.

'Tim Magee, and his colleagues at CSDi, are to be commended for producing a book which should change the way development is practiced, and so directly contribute to the improvement of millions of lives around the world.'

Howard White, Executive Director of the International Initiative for Impact Evaluation (3ie), USA

'A fascinating and informative guide to a subject of growing international importance. Tim Magee skilfully explains ways to combine external expertise and local perspectives on adaptation to climate change. This useful book should be read by development practitioners as well as students of climate change policy and international development.'

Tim Forsyth, London School of Economics and Political Science, UK

'This is a most-awaited book for development practitioners who are increasingly confronted with the challenge of addressing climate risks in designing and implementing programmes and projects. This book will help them to do just that in a way that places the interest of communities at the heart of the process.'

Kareff Rafisura, Climate Risk Management Practitioner, Ghana

'This book provides an insightful and comprehensive field guide to community-based adaptation. Magee brings together an impressive range of tools, resources and case examples in a clear and systematic step-by-step guide, while ensuring that the concerns of local people are kept at the centre of the analysis. This book is a timely and welcome addition to the literature, and will be useful to experienced practitioners as well as newcomers to CBA.'

Lars Otto Naes, Institute of Development Studies, UK

A Field Guide to Community Based Adaptation

Tim Magee

With a foreword by Howard White

Routledge
Taylor & Francis Group
LONDON AND NEW YORK

earthscan
from Routledge

First published 2013
by Routledge
2 Park Square, Milton Park, Abingdon, Oxon, OX14 4RN

Simultaneously published in the USA and Canada
by Routledge
711 Third Avenue, New York, NY 10017

Routledge is an imprint of the Taylor & Francis Group, an informa business

British Library Cataloguing in Publication Data
A catalogue record for this book is available from the British Library

Library of Congress Cataloging-in-Publication Data
Magee, Tim.
A field guide to community based adaptation / Tim Magee.
p. cm.
Includes bibliographical references and index.
1. Climatic changes—Economic aspects—United States. 2. Community development—United States. I. Title.
QC903.M34 2013
363.738'74560973—dc23 2012025388

ISBN13: 978-0-415-51929-8 (hbk)
ISBN13: 978-0-415-51930-4 (pbk)
ISBN13: 978-0-203-08255-3 (ebk)

Typeset in Times New Roman
by RefineCatch Ltd, Bungay, Suffolk
Illustrations by Gary Holmes

Contents

List of Tables

Foreword

Some years ago I was evaluating community-driven development programmes in rural Zambia. Many communities chose to use funds to build, or usually re-build, school classroom blocks. The approach of a well-known international NGO was to hand the money over to the communities and say 'okay, go build the school'.

Another programme in use at the time, the social fund project, took a different approach. Communities took part in construction, supplying building sand and moulding bricks. But the actual construction was done by a trained contractor following plans supplied by the social fund, with a budget based on regional unit costs. The social fund-supported schools, clean blue and white brick structures, were readily recognizable as I travelled around the country. The schools built by communities using the NGO's approach had fallen down, or remained uncompleted as the budget had been poorly prepared or funds misused.

The lesson from that story is that community-driven approaches need to be implemented within an institutional framework that combines community-based priorities and skills with outside expertise and skills. It is a mistake to think that a community driven approach implies that communities have within them all the skills and knowledge necessary to identify, develop and implement development projects. This process needs to be carefully facilitated, applying international knowledge of what works to local conditions, and providing an institutional framework within which project activities take place.

For those wanting to apply this approach Tim Magee's new book provides a step-by-step guide.

Evidence-based development is a new buzz-word in development. Many agencies are struggling with what this means. Tim and his colleagues at CSDi have taken this approach right down to the grassroots. They are not discussing what it means but getting out there and doing it.

My own organization, the International Initiative for Impact Evaluation, promotes the use of evidence about what works and why amongst developing country governments and international development agencies. Tim's work in Guatemala has taken this approach to the field, feeding international evidence into locally-based solutions. The on-line field guide, on which this book is based, has taken the approach to over 100 countries around the world.

I outline here what I see as the key strengths of CSDi's approach:

- *Participatory*: Community Based Adaptation (CBA) is based on the well-established principle of utilizing local participatory processes to ensure that the chosen project meets a local priority need. Sustainability is thus more likely because community engagement builds ownership, and their involvement in project identification and implementation encourages the use of appropriate technologies which can be maintained locally.

- *Solution-based*: The priority need is first identified and then the most appropriate project identified. This is the right way round. It is often done the wrong way round. That is, many interventions are decided on the basis that 'this is what we have always done' rather than an assessment of the most appropriate intervention to meet the chosen objective. Solutions are based on combining local priorities with an application of international perspectives to local conditions, to produce a participatory but rigorous analysis of local challenges from climate change.
- *Evidence-based*: Draws on rigorous international evidence both to analyse the problem and to know what works. The Field Guide provides concrete examples of the types of evidence to be used, where to find it and how to use it. The salient point is made that various levels of 'knowledge brokers' may be necessary to mediate this knowledge to the field level. Using good evidence is necessary to protect against wrong-headed if well-meaning outsiders, and to provide the framework for utilizing local knowledge and skills. Technology is appropriate because it meets a local need, and can be used on a sustainable basis given local skills and resources. Appropriate does not mean restricting the possibility set to what communities have already or can think up themselves.
- *Theory-based*: The solution is based on an analysis of the problem, the causes of that problem, and so how to address it. The theory-based approach allows identification of critical assumptions, and so conditions that need be in place and actions to be taken for the project to succeed. As stressed above, the approach draws on international research applied in the context of a strong understanding of local conditions.

CBA is orientated toward adaptation projects. But as stated in the book, adaptation projects are not so different from many other general development projects. So the appeal of this book is far wider than those interested in the challenge of adapting to climate change at local level. The book will also appeal to anyone committed to community-based or community-driven development. Hence the Field Guide should command wide readership and use throughout the development community, from small local NGOs to the largest multilateral development agencies.

Tim Magee, and his colleagues at CSDi, are to be commended for producing a book which should change the way development is practised, and so directly contribute to the improvement of millions of lives around the world.

Howard White
Executive Director, 3ie

Introduction

Why Write This Book?

Two of the most important groups of people in community based adaptation are field staff—the people on-the-ground doing the work—and the communities that they work with. This book has been written for you—a development practitioner who wants to begin working today, with your communities, to successfully adapt to the climate challenges they are facing.

The best-intended generosity of donors and the most carefully researched development program won't have the intended impact if trained, effective field staff are not on site to implement projects—and if well-prepared communities are not ready to take over when the non-governmental organization (NGO) staff depart. Without the cooperation of these two groups in project design and management, project outcomes run the risk of coming to an abrupt halt at grant's end—undermining the long-term impact that all parties had worked so hard for.

The world's poor will be the most critically affected by a changing climate. These human beings have been struggling for decades with poverty, malnutrition, poor health, and a lack of access to education. Development professionals are beginning to realize that, for many communities, their plight may not be improving rapidly enough to reach the UN's Millennium Development Goals by the targeted date. If experienced development organizations are finding it difficult to solve decades-old development problems, how will they additionally solve new challenges driven by a changing climate?

An underlying cause—with a veiled solution—may be that development has not always included communities in the process of assessing need, designing project activities, having a stake in project management, and taking over project stewardship at grant's end. Cutting-edge development now sees this community participation as paramount for maintaining the positive outcomes that contribute to long-term impact. This is the beauty of community based adaptation (CBA): it both improves development results by fully engaging community members as partners and owners—and at the same time increases the community's resilience to a changing climate through sound adaptation practices.

A second underlying reason for a lack of forward motion in development may be that many development practitioners do not have access to the field tools, sound information, and training necessary to improve project impact. Without effective tools and methodologies, how can these practitioners learn to design projects that are both sustainable and impact-oriented? How will they learn to introduce CBA activities into traditional development projects when resources are already stretched thin?

The goal of this book is to increase the effectiveness of community based adaptation so that many, many more people worldwide can enjoy increased resilience to climate change challenges. This books hopes to do this in two ways. One: by leverage. If this field guide can help put sound tools and techniques into the hands of in-country staff, the staff will be better prepared to design and implement impact-oriented projects for increasingly greater numbers of people. Two: encouraging in-country staff to engage community members in project design and implementation will increase the likelihood that positive project outcomes will be sustained by communities for the long term.

After observing the challenges that field staff face with a relatively small arsenal of tools and techniques I decided in 2008 to found the Center for Sustainable Development—a US registered charity. The Center specializes in providing accessible, sound, evidence-based information, tools and training for humanitarian development professionals worldwide. The Center is firmly committed to proven, results-based solutions to end suffering and poverty. The Center's goal is to spread these solutions across the globe through online field guides and interactive online training. My thanks to the many friends, supporters, and students of the Center who have made this possible. This book, the *Field Guide to Community Based Adaptation*, takes this dissemination of information to a new level.

What the Book Will Do for You

Where does one turn to for information on designing, launching, and managing CBA projects? For many staff members it may feel overwhelming to launch an adaptation project. Where do you begin? What practical tools are working today? How do you convey climate change concepts to a community? Together, through this book, we will explore the full process of developing CBA programs.

This field guide provides a system for field staff to use in developing successful projects that can be co-managed and sustained by communities, and includes the tools needed to do the following:

- assess and organize information about climate challenges and vulnerabilities that communities face;
- work side-by-side with communities to develop long-term, sustainable adaptation programs;
- link development, disaster-risk reduction, and adaptation activities into sound projects;
- empower communities to take full charge of programs once the programs are up and running.

Why is community engagement in project design and implementation important? A criticism of the traditional project cycle is that when an NGO completes its two-year project, they leave their community at the helm of project management without sufficient training and technical support—and perhaps even without much interest in the project. For example, how many communities have you been to and seen two-year-old water projects that no longer function?

Unlike a childhood vaccination program, where once vaccinated, neither the child nor the parents need to oversee project continuation, CBA projects are for the long term: decades. If they are to be successful projects, the behavioral changes, forest management schemes, or the water harvesting systems you promote will need to be maintained indefinitely—or the positive results will wane. Community based adaptation ensures that communities are engaged from the beginning of the project concept and are involved in each important step of the process. Community engagement creates the caring ownership that sustains these important adaptation components for long-term sustainability.

A Field Course in a Book

The book is arranged in a progression where each chapter represents one sequential step in the process of project development, including problem assessment, project design, project launch, and community adoption. Each chapter has the tools needed by field staff to complete that chapter's step in developing a real, on-the-ground project. So that you can stay on track, the field course follows one consistent, example project as it grows and develops during the step-by-step field assignments.

The field activities in each of the chapters are designed to be accessible by people at different levels of your organization—and to be able to be started quickly. The low cost/no cost activities are simple enough for implementation by field staff with basic skills, and for adoption by community members with basic capacities for sustaining activities. You can get started empowering your community today with this very powerful set of tools. For those of you wanting to explore individual techniques in greater depth, each chapter provides recommended resources for further study.

There are four parts in this book. Part I deals with discovering and defining the full climate change context that a community of people find themselves in. Part II develops a full project, complete with solution-oriented activities, management tools such as log frames, and an introductory fact sheet for presentation to donors. Part III launches and implements the project—and hands it over to the community as part of your exit strategy. Finally, Part IV is a collection of how-to field guides on specific activities for you to use in your project.

CBA Project Activities

There is some confusion over what adaptation activities are. Many people expect a palette of brand-new technologies to begin arriving any day now: adaptation silver bullets. But problems caused by a changing climate will likely pose a risk for livelihood, health, food security, and access to water—many of the same things that traditional development projects struggle with. CBA project activities will in many cases be the same activities used in a development project—but simply used to solve a different problem: challenges driven or intensified by a changing climate.

The book will therefore promote cross-cutting initiatives that combine traditional development, disaster-risk reduction (DRR), and CBA activities. In a cross-cutting—or mainstreamed project—an activity can be incorporated specifically to address a traditional development challenge, a challenge related to climate change—or perhaps a solution specifically designed to reduce risk associated with disasters. Combined, these interrelated activities work in support of each other and facilitate community members in developing resilience, safety, and prosperity. For example, by adopting the use of drought-resistant crops, farmers can adapt to a drier climate. Through improved health, nutrition, and preparedness for disaster, communities will reduce their vulnerabilities and increase their resilience to climate change impact.

The Steps

How do you know if your community project is linked to climate change? What practical, CBA tools, solutions, and activities are available today that you can include in your project? In order to address these questions, the book offers two approaches. The first approach is to use the step-by-step field assignments (one per chapter) that show how to design, launch, and manage a sustainable CBA project. The second approach is a collection of ten hands-on field guides about individual CBA activities for readers to use in solving classic climate change challenges.

In Chapter 1, you will learn how to assess the challenges that a community faces. In Chapter 2, explore their understanding of how the climate is changing around communities, and learn how to uncover the adaptive solutions which they have developed on their own. In Chapter 3, you will research scientific information about climate change for your specific location. This will help you determine which are the most likely climate change (CC) scenarios that are unfolding now—and what will play out in the future.

In Chapter 4, you will compare your community's local knowledge with scientific knowledge, verify that the community's challenges are linked to climate change, pinpoint underlying causes, and put together a clear definition of the local climate change context. Chapter 5 begins with the selection of specific field activities to use in strengthening community resilience. Before finalizing a project outline, you will confer with the community to get their feedback and input on the emerging plan.

In Chapter 6, you will develop the management and funding tools that you will need to launch the project. Your project outline will be transformed into a logical framework complete with a monitoring and evaluation plan, and expanded with a detailed budget, schedule, and a concise project fact sheet—all for presentation to a donor. The development of each of these documents will be taken in simple steps using downloadable templates and examples.

Chapter 7 quick-starts the section on project implementation where you will organize a community based project management team. In Chapter 8, you will initiate the process of handing the project over to the community as part of your exit strategy. This will include planning for long-term management, and developing a plan for long-term technical support. You will develop adaptation capacity building workshops for community members, and officially launch the project in partnership with the community.

Chapter 9 will help you design a post-project participatory monitoring and evaluation plan so that the community can continue to evaluate their project, learn from their project, and fine-tune activities in an effort to stay on course.

Chapter 10 is a collection of hands-on field guides on CBA activities that have been compiled for use in your project. These field guides include a snapshot of the activity, illustrated workshop handouts, and participatory workshop lesson plans. Ten examples of project activities have been chosen that address adaptation challenges that communities face in the areas of water, food security, agriculture, disaster-risk reduction, and livelihood diversification. They very likely represent solution-based activities that you could use as-is in designing and implementing your project.

The field guides chosen for this chapter represent ten of the most universal challenges that communities face in building resilience and in adapting to a changing climate:

- Participatory Community Needs Assessments;
- Participatory Capacity and Vulnerability Assessments;
- Preparing Family Garden Beds and Planting Seeds;
- Participatory Mapping of Soil and Water Resources;
- Soil Restoration and Conservation for Smallholder Farmers;
- Agricultural Soil and Water Management for Sloping Land;
- Household Rooftop Rainwater Harvesting;
- Community-Level Water Harvesting;

- Developing a Community Based Disaster Risk Reduction Plan;
- Diversifying Livelihoods through Market Links.

The Appendix to this book gives the background to the course and details the successful project activities undertaken by the students, and the list of international organizations that have benefited from our courses.

I hope that the book will be useful and enjoyable, and will help you in assisting a greater number of people adapt to a changing climate, break out of the cycle of poverty, contribute to the development of their communities, and lead fulfilled, meaningful lives.

Part I

Local Context

The aim of this book is to help you develop, fund, manage, and hand over to a community a sustainable, community based adaptation project. It is a field course in a book that provides you with a series of concrete, practical steps to use in working with a community, with technical experts, with a donor, and with your team to develop the project. You will create a personalized template during each one of these steps that you and your organization can use for future projects—increasing your efficiency in project development.

Part I has four chapters. In Chapter 1, you're going to start working immediately with a community by using participatory tools to discover their perception of their needs. You will also begin the process of organizing community-identified need into a very simple project outline.

In Chapter 2, you will then dig more deeply into the community's vulnerabilities to climate change, gain a better understanding of their local knowledge of climate change, and learn about their coping strategies for adapting to climate change. Next, in Chapter 3, you will analyze scientific and meteorological information that is specific to your community's location.

In Chapter 4, you will compare local need, local climate knowledge, and scientific climate information, and then organize your findings into a statement describing the specific context of your community's climate change challenges. You will then expand the project outline developed in Chapter 1 with this newly compiled information.

1 Community

Local Needs

Participatory Needs Assessments

You're going to start working immediately with a community by facilitating a participatory needs assessment. The assessment will be done in a community workshop that will take approximately four hours. In this workshop you will work with the community to explore their thoughts on their needs, wants, problems, and challenges. You're going to provide attendees with communication tools designed to illuminate and capture their knowledge about their community and about their needs. Given the right tools, most communities are fully capable of assessing their needs and designing adaptive programming.

Why Have a Participatory Needs Assessment?

1. Community members may have a greater depth of knowledge about their problems, vulnerabilities and coping strategies than you do, and so will be better able to identify important and underlying causes for the challenges they face.
2. If they are fully engaged in project development and feel their voice has been heard, they will have a sense of project ownership. Their project will then be on the road to sustainability because in essence it is their project—they own it.

Chapter 1 will help you accomplish four steps in the development of your project. You will do the following:

1. learn about communities, challenges and sustainability;
2. learn how to develop a project based upon a participatory needs assessment;
3. facilitate a community based needs assessment;
4. create a simple project outline.

What you'll need:

- access to a community;
- representative community members to participate in a workshop;
- workshop materials, snacks, and drinks.

Timeframe:

- 7 hours:
  - 2-hour workshop preparation;
  - 4-hour workshop;
  - 1-hour organizing workshop results into a simple project outline.

Initiating a Project Outline

Your last step is to organize the community's challenges into a very simple project outline capturing the visible problems, their underlying causes, and their long-term negative impacts.

Course Project Examples

The example field assignments provided in each chapter represent a single, unique course project that grows and develops chapter by chapter throughout the book. This will give you an unfolding, consistent project to compare your project with, to measure progress from one step to the next. Example field assignments can be downloaded and used as a template for completing an assignment. They represent what the assignment should look like and what information it should contain. To use the example templates, simply download them from the book's webpage (TimMagee.net/field-guide-to-cba/) and use them to complete your assignment.

Introduction to Chapter Resources

At the end of each chapter is a section called Chapter Resources which will contain the following three resources:

1 *Suggested homework assignment*. If this book is being used in conjunction with a course, these are suggestions for the homework to be turned in upon completion of the assignment. The suggested activities are parallel to the assignment instructions in each chapter and to the example of the completed field assignment provided at the end of each chapter. If this book is being used to develop a field project, the homework examples can be downloaded and used to simplify project design and development. One assignment builds upon the next, resulting in a complete set of project management tools. These completed assignments can be used both in managing your project, and for presentation to a donor.
2 *Course downloads*. This book has a companion webpage (TimMagee.net/field-guide-to-cba/) where you can download the course resources, and the examples of the completed field assignments that have been developed for each chapter. These downloads will give you the ability to adapt and modify materials to better fit your organizational needs and the needs of the community that you are working with.
3 *Recommended resources*. These are a collection of useful handbooks, manuals, and papers that are available online from the organizations that produced them.

Field Assignment 1 Step 1: Communities, Challenges, and Sustainability

In this chapter you will begin developing a community based adaptation project right away. How will developing a community based project differ from developing a traditional project?

The techniques used in developing a community based adaptation (CBA) to climate change project are to engage a community at the very beginning in project design, to work with them to develop appropriate project activities, and to foster their development of their skills sets that will allow them to be the long-term stewards of project activities and outcomes. The community will need to continue project activities long after you are gone if they are going to adapt successfully to long-term climate change.

> Community-based adaptation to climate change is a community led process, based on community's priorities, needs, knowledge, and capacities, which should empower people to plan for and cope with the impacts of climate change.
>
> It must draw on the knowledge and priorities of local people, build on their capacities, and empower them to make changes themselves.
>
> (Reid *et al.* 2009: 13)

What is a Community?

A community is a group of people that you serve. A community could be all of the members of a remote village—or more specialized communities such as fishermen on a small island, pastoralists on the savanna, members of a farmers association, or a group of urban women who want to increase food security for their families. In the North, a community could be people visiting a food bank, new immigrants arriving in your city, teenagers who visit a teen drop-in center, or residents of a retirement home.

What is a Project?

A project is a group of activities that you have theorized will provide long-term, sustainable solutions to community-identified need. A project should be carefully designed to be sustainable, fundable, manageable, and successful at achieving carefully defined outputs, outcomes, and long-term impact. It should also be designed—from the beginning—to be taken over and continued by the community at the end of your project cycle.

What are the Challenges that are Linked to Climate Change?

These challenges are as diverse as traditional development challenges and can include health, water, food security, migration, livelihood—and disaster preparedness. When comparing traditional development field activities with adaptation field activities, the activities can look very similar—in fact, they can even be the same activities—just used to address different underlying causes to the challenges.

Climate change challenges fall into three categories:

1. highly specific challenges caused by a changing climate such as the shift in the beginning of the rainy season (affecting the farmers' planting cycle) that need to be addressed using specialized adaptation activities;
2. traditional development challenges such as a decrease in food security, making communities more vulnerable to a changing climate. These can be addressed with traditional development techniques to increase community resilience to climate change;
3. climate-related disasters such as extreme weather events that can be addressed with traditional disaster risk reduction techniques to increase community resilience.

> Community-based Adaptation brings together those working in the fields of disaster risk reduction (DRR), community development, and climate change science. Community-based Adaptation draws on participatory approaches and methods developed in both disaster risk reduction and community development work.
>
> CBA needs to start with communities' expressed needs and perceptions, and have poverty reduction and livelihood benefits, as well as reducing vulnerability to climate change and disasters. In practice, CBA projects look very like "development as usual" and it is difficult to distinguish the additional "adaptation components."
>
> (ibid.: 4)

It is important in the beginning of project design to determine whether the problem that the community is facing is indeed related to climate change or to a traditional development cause. The reason is that an activity that might work to solve a traditional development problem might not work for a problem that is linked to climate change—even though the visible problem is the same. Different underlying causes will require different solutions.

A simplified example could be that a village spring has dried up (visible problem). Is this due to the fact that there is a climate change-induced long-term drought (i.e. climate change is the underlying cause)? Or is it due to the fact that the hills behind the village have been deforested and rainwater simply runs off, therefore it is no longer able to infiltrate slowly into the soil and recharge the spring (i.e. traditional development is the underlying cause)? These are two very different underlying causes to the same visible problem (village spring has dried up).

A watershed restoration that includes reforesting the hills could mitigate a runoff problem if there is a rainy season, but might not solve the water shortage problem if indeed the spring has dried up because of a long-term drought. Consequently, underlying causes must be carefully evaluated and verified.

What is Sustainability?

For the purposes of this field guide, we are going to look at two different definitions of sustainability—both extremely important.

The first is the beautifully universal, famous quote from the UN Brundtland Commission in 1987:

> Sustainable development, which implies meeting the needs of the present without compromising the ability of future generations to meet their own needs.
>
> (United Nations General Assembly 1987)

Communities and Sustainability

A second definition is perhaps less eloquent than the Brundtland Commission's definition, but addresses a slightly different challenge. Once a successful activity has been launched within a community to increase a community's resilience to climate change, they need to sustain that activity for as long as the challenge still exists. In the case of climate change, this could be for decades: climate change is for decades. Community members will need the capabilities and resources to continue, sustain, and maintain long-term activities.

Frequently in development, we may think about a project as having a relatively short duration that may not need to be actively maintained by the community after the NGO has left: a childhood vaccination program could be an example of this. However, if your project involves techniques such as soil conservation, water conservation, the restoration of a watershed, or improved agricultural practices, these activities will need to be continued and maintained by the community long after you—and your NGO—have departed. Project activities, therefore, need to be designed from the very beginning to be able to be sustained by the community.

For example, if unpredictable heavy rainfall floods farm fields and damages crops, a community may choose to develop a water management plan which includes building channels to divert flood water away from the farm fields. For the solution to work indefinitely, routine maintenance, repair, and improvements to the channels will need to be sustained indefinitely. Why would community members do that? If you have done your job, they will feel that it is their project: they will have a sense of caring ownership of the project.

Consequently, sustainability implies that you need to begin developing a sense of community ownership from the very beginning. Ideally, you would want your community to view the project you are developing as having been their idea—and you're simply a short-term consultant who's coming in to give them a spot of expertise. So in this first chapter, you are going to begin engaging with the community and helping them to develop this sense of project ownership right away.

How to Develop a Sustainable Project

1 Develop a project that the community wants and feels that they own.
2 Have a solid definition of the impact that you are hoping to achieve.
3 Use solution-oriented activities that have shown evidence of having worked to solve your context-specific challenges.
4 Ensure that your solutions will not create secondary challenges or future problems.
5 Know how to design, fund, launch, manage, and hand over a solution-oriented, successful project.

Beginning in this chapter you will begin walking through a step-by-step process to develop a real project. Each chapter will have a field assignment; portions of the field assignments will be conducted in the field with your community, other activities will continue project development through research and investigation.

Field Assignment 1 Step 2: Developing a Project Based upon a Participatory Needs Assessment

The goal of the first field assignment is to determine community need from the vantage of the community members that you serve. Why is this important to do? As donors or NGOs, and as human beings, we are all guilty of assuming that we know what is best. But what is best for us may not be what is best for another person from another culture. A community's view of their needs may include information that they have and that we know nothing about. We also need to acknowledge their perception of their needs and to work to address challenges that the community has prioritized; this increases community ownership.

In this chapter you will be designing a participatory community needs assessment workshop. You are going to provide attendees with communication tools designed to illuminate and capture their knowledge about their community and about their needs. This will help you to better understand which pressing needs they are facing and which ones are the most important to them.

Given the right tools, most communities are fully capable of assessing their needs and designing adaptive programming. Your job, as a development practitioner, is to walk hand-in-hand with the community accompanying them in a two-way learning adventure, facilitating leaps of insight and understanding, and providing specialized skills when they need it. Your job is not to be the leader: your job is to empower leaders.

Your Participatory Needs Assessment

There are several very positive reasons for encouraging your community to participate in the process of defining and prioritizing their needs:

1 Community members may have a greater depth of knowledge about their problems than you do, and so will be better able to identify important and underlying causes for the challenges they face.
2 If they are engaged in the process of needs identification, and feel their voice has been heard, then they will have a sense of ownership for the process and the outcomes; this leads to long-term project sustainability. This ownership manifests itself as the community's demand for the services that your organization will provide that are necessary to launch and manage their project.
3 Working with a community to address their needs will develop trust on their part in working with your organization on future projects or activities.

First, you need to develop a relationship within a community

Let's say that the community you wish to work with is a small rural village that is new to your organization. In approaching a new community for the first time, it is best to approach community leaders initially, let them know the purpose of your visit, and help them to understand the importance of your work.

It is a good idea to have introductory materials about your organization, the kind of work that you do, and the positive results that you have achieved in other locations in the past. You might also want to explain a bit about your method of engaging community members early on in the process—with the first step being a participatory needs assessment. Let them know that the purpose of the needs assessment is to get to know the people better, and to better understand their lives, needs, and challenges.

Be careful: these village leaders may feel that they can give you all of the information that you require. If that is the case, suggest two participatory meetings: one with them, and a second one with community members in the greatest need.

With the support of the leaders, you will be able to meet community members interested in participating in your project—and in the initial needs assessment. Ask if they can help set up a 4-hour meeting with 10 or 12 people. Suggest that you would like to meet with community members who represent the ultimate beneficiaries. The size of the group for your first assessment should not exceed 12 people.

Communities are diverse and you need to be sure that you are working with a representative example of its members. Each sub-group of community members will have their own set of needs; some members may even be self-serving. Consequently, you will need to choose which groups will be the most representational of the overall community need. You might even need to do two assessments. For example, one could be conducted only with women so they can feel free to speak, and a second assessment with men in the community.

Box 1.1

Organizational Mission Versus Community Need

There are many stakeholders in the development process: your organization, your donor, the local government, the village leaders, and the community members themselves. Each stakeholder has their own mission. You can begin to see that with all of the different stakeholders involved, it can be difficult to assess and prioritize real community-identified need.

For a true participatory assessment, you will need to exercise some critical self-evaluation. In an ideal setting, you would start your project design by entering a new village, developing relationships, and then engaging community members in an open needs assessment process. However, even your own organization can complicate this process by coloring it with its own mission focus. For example, here are some problems and potential solutions:

- Your organization has a specialty. Let's say that you focus on agriculture. How do you balance your organization's specialization with needs defined by the community that are not agricultural in nature?

Potential solutions: You could partner with another NGO to address community needs that do not fit your specialization. You could decide to expand your organization's capabilities and seek training in a new specialty. You could hire a specialist to run that aspect of the project.

- What if the community comes up with a top-priority need that you don't think is important, or don't think will do any good?

Potential solution: You will need to weigh the costs and time investment of implementing their priority against building good will and trust between your organization and the community. Could their priority be included in a project component as a sub-component?

- What if you are already working in a community, and have an established relationship and an ongoing project with them?

Potential solution: A community needs assessment at this point may be an excellent idea. It can give your organization quality feedback about your programming. Remember, this is about long-term sustainability; if your community is not buying into your current programming, the project might not last very long after you leave. A needs assessment will offer your organization two things: feedback for fine-tuning current project activities, and new ideas for the next project and funding cycle.

Field Assignment 1 Step 3: Facilitating a Participatory Needs Assessment

Activity 1: Organizing a Needs Assessment Workshop

This activity is expanded upon in Chapter 10, Field Guide 10.1. You can also download a needs assessment lesson plan and the Example of Field Assignment 1 from Chapter 10.

Begin by setting up a 4-hour meeting with 10–12 community members. To facilitate your learning experience, smaller groups will be easier to manage during this assignment; you can expand upon the assessment at a future date.

Arrange to meet with community members who represent the ultimate beneficiaries (mothers, fathers, families, farmers, weavers, whoever best describes the community you are working with); try to avoid basing your entire assessment on a meeting with people in higher positions, for example, mayors or village leaders. It is important that women and marginalized community members also have a voice in the process. You may need to have several meetings with the different groups so that individuals will feel most comfortable expressing themselves.

Begin organizing the workshop itself at least two weeks in advance. Make sure that you have all your materials, like large sheets of paper, and pens and markers for drawings. Since this is a four-hour workshop, you may also need to plan for snacks and drinks.

First, download the lesson plan from Chapter 1, Resources: it explains the entire workshop in a step-by-step fashion, or read through the lesson plan in advance (Chapter 10, Field Guide 1). Review the exercises with your team and make simple adaptations to the exercises so they are appropriate for your community's situation. You may choose to produce an illustrated handout or poster for the workshop, especially if some participants cannot read. Role-play the exercises with your colleagues so that you will be better prepared when you present the workshop, and so you can discover if there are any cultural or linguistic problems.

Have two to three colleagues accompany you to help. This will be especially useful if you decide to break the participants down into sub-groups (for example, men and women, or teenagers and parents). If you are considering providing snacks, drinks or a lunch, put someone in charge so that you are not distracted with the details and are free to focus completely on facilitating the workshop. Put someone in charge of taking photos. Photos are a great reminder of what you did, they are excellent for training your teammates, and they can be used to promote your organization.

Box 1.2

Participatory Needs Assessment Overview

In community based adaptation, the development activities you want to work with are grassroots solutions for addressing community-identified need. In this first community based workshop you are going to use a Participatory Learning and Action (PLA) ranking tool. In the process, workshop participants voice different problems, challenges, and needs they experience in the community—and then vote on them with voting tokens (small stones or beans) to prioritize them. You will use drawings to illustrate community-identified needs so that illiterate community members can equally participate in this process.

How It Works

After initial rapport building with the group, explain that the purpose of the activities is to understand and learn about their community from their perspective. Ask the group to imagine and discuss the problems and needs that are faced by the community as a whole.

As each need is identified by a community member, begin making simple illustrations that represent the challenges they describe on notebook-sized sheets of paper (you can bring a selection of typical drawings to reduce time spent drawing during the workshop). An example could be that if there is a housing shortage, draw a little house. After the group has come up with a complete set of needs and challenges, arrange the different illustrations side-by-side into a rectangle on the ground or on a table.

Have everyone leave the workshop area. Give each one of the participants voting tokens—10 or 15 slips of paper, or beans, or grains of corn. For privacy during voting, only one person should go into the workshop area at a time to vote. They should select the needs which they feel as an individual are the most important. It is their decision if they want to put all 10 tokens on one drawing or if they want to distribute them around several different challenges.

When the participants have finished voting, count the total tokens on each drawing and write up a prioritized list ordered by the number of votes each problem received, with the need that received the most votes at the top.

Give the participants a break so that you can take a few minutes alone with the list and draw a two-column matrix on a sheet of newsprint. In the left column, write down the individual needs in their prioritized order (or draw little pictures again) with the needs receiving the highest votes at the top of the list. In the right column, write the number of votes each one received. There is an example illustration of this matrix in Chapter 10, Field Guide 1.

Post the matrix where everyone can see it. This is a good time for the participants to have an open discussion about the results of the vote. Plus, if there are any unrelated needs competing for the highest position, it would be a good idea to let the participants choose which one they feel is the most important. For example, there might be a health-related challenge near the top and a microenterprise challenge near the top as well. You can ask the community which project they would like to start with first. This will help keep your project simple and not put you in the position of having to manage two dissimilar programs at the same time.

It is very likely that the list will be a disorganized mixture of needs, challenges, underlying causes, and grievances. Work with the group to connect needs and challenges to their underlying causes on the matrix so that they can see the relationship. If the matrix does not have any underlying causes, this would be a good time to ask the participants what they feel the causes of the top priority challenges might be. It is likely that they have more background information about a problem than you do, so this can be quite helpful.

Conclude the meeting by summarizing the two or three challenges that the community placed as their highest priorities. Ask for participant feedback of your summary for verification. Use your best facilitation skills to make sure that no one has any questions or additional comments.

This summary and the lesson plan for conducting this workshop can be downloaded from the book's website: visit Chapter 1, Resources. They are also available in Chapter 10, Field Guide 1.

Field Assignment 1 Step 4: Creating a Simple Project Outline

Activity 2: Organizing the Results of the Participatory Needs Assessment and Development of a Simple Project Outline

Go to Chapter 1, Resources, to download a template called "Example of Field Assignment 1." It will be helpful in analyzing and organizing the results of your needs assessment. A simplified version of this template is given in Text Box 1.4.

The step-by-step process used in this book will work best if you work with a very simple initial project. You will learn the techniques more quickly since your learning process will not be complicated by the intricacies of a larger-scale project. Also take into consideration that frequently a simple project can be more easily communicated to staff members, community members, and to donors.

The first thing that I did in my example was to write a very short, one-paragraph narrative summary about the community, about the participants whom I worked with in the assessment, and about the results. Second, I made a simple matrix showing the results of the prioritized needs assessment. I arranged it so that the highest priority is at the top and shows the number of votes that each challenge received.

Next, I created a very simple project outline with the visible, prioritized problems listed at the top of the outline, their underlying causes listed beneath them, followed by the long-term negative impacts created by the problems. I then collapsed the outline into a short paragraph which became the problem statement of the project. The problem statement is absolutely parallel to the project outline: I simply copied the individual components of the outline and pasted them into the form of a paragraph.

Here's the process that I used. Take the project challenge that you agreed to work on with the community at the end of the assessment, and using the Example of Field Assignment 1 as a template, write an outline of the following:

- a problem or two;
- an underlying cause or two;
- some of the long-term negative impacts that will result from the problem;
- a short paragraph (Problem Statement) that is nothing more than the combination of the problems/causes/negative impacts from the simple outline. The problem statement is not an introduction to a proposal, nor a paragraph of background information. It is simply the problem, the underlying causes, and the negative impacts copied and pasted together in order.

You can simplify this process by downloading the Example of Field Assignment 1 and using it to create your own project outline.

Unraveling the Mixture of Needs, Problems, and Causes

In looking at my prioritized list of community-identified needs, I needed to make some decisions about how to organize the information into an outline. See Text Box 1.4.

Since stunting and poor school attendance are long-term negative impacts, I put them in the negative impact statement and added their consequences to make it a compelling statement.

However, the health and nutrition challenges (sick, malnourished children) are human scale, visible, solvable, and trackable—so I made them the over-arching problems—and put

Box 1.3

Overview of Problems, Underlying Causes, and Negative Impacts

When you meet with your community to do the needs assessment, they will present you with a mixture of needs, problems, underlying causes, grievances, and negative impacts. Your job as a facilitator is to encourage them to say everything that is on their mind. Their votes will prioritize the two or three things that are the most important to them, so this will simplify your job of organizing an outline.

Your job is to be a facilitator and interpreter, and to sort their array of challenges into three things:

1 one or two important problems that they have prioritized;
2 one or two underlying causes of those problems;
3 the long-term negative impacts that the problems cause.

Problems, for the purposes of this book, are the visible and compelling elements of the needs assessment. These are the things that human beings can relate to. For example, sick little children or malnourished little children are visible, compelling problems. You can see them, you can feel their pain and suffering. However, if the community lists, for example, contaminated water—realize that this is not a compelling problem—it is a cause of a problem. Look for the visible, compelling problem that contaminated water causes.

Underlying causes in this book are the project outline components that are the causes of the highest priority problems that your community has identified. Contaminated water (from the example above) and a lack of knowledge of health and hygiene (safe water storage, hand washing, kitchen hygiene) are good examples of the underlying causes that lead to sick children. An overall shortage of food is an underlying cause of malnourished children.

Negative impacts are the long-term negative results of the problem. Sick children do not function well in school, have trouble gaining an education, and may therefore be unable to lead the prosperous, meaningful, productive lives that they need as adults to leave the cycle of poverty. Negative impacts are long-term outcomes—5–15 years away. So your project is going to address the immediate problem (sick children) in order to reach the long-term goal (positive impact) of healthy, productive, well-educated community members. Follow this process to organize the needs assessment into a project outline:

1 Decide what the visible, compelling problem is (if all that the community comes up with are causes, ask them about what the highest priority problem is or use your observational skills).
2 Determine what the underlying cause is for that problem.
3 Determine what the long-term negative impacts are.
4 Fit them into an outline template like the one in Example of Field Assignment 1.

Keep your problem/cause/impact outline short and simple. If your community raises a number of challenges to be addressed, you can return and develop separate projects for those challenges at a later date.

them at the top of the outline. Then I was able to choose underlying causes for each problem from the rest of the community's list.

Go to Box 1.4 to see what the matrix and project outline should look like, or download the Example of Field Assignment 1.

Box 1.4

Example of Field Assignment 1

What's the real problem?

This example project is based upon four communities in the highlands of Guatemala. Some 33 women from an association of women weavers in four remote villages in Comalapa, Guatemala, participated in a needs assessment. Each woman was able to voice needs and then vote on them with 10 beans. These 33 women were representative of 100 families—25 in each of the four villages—that my NGO hopes to work with on a new project. The needs they described were improved health and development for their children, better school participation for their children, and increased food security, and were expressed as the following problems (Table 1.1).

Table 1.1 The full list of needs/problems and the vote results

The full list of needs/problems	*Votes out of 330*
Stunting: lack of physical and mental development in children caused by:	81
Chronic diarrhoea in children	
Chronic under-nutrition	
Shortage of family food for four months preceding the corn harvest	74
Increasingly reduced crop harvests	70
The need for more weaving work and more income	50
Poor school attendance and performance for their children	35
Micro-businesses affected by women's inability to read and write	20

Since I wanted to develop a simple project, I chose only the top two visible and compelling problems and only two underlying causes. We decided to design a project that would address food security, health, hygiene, and nutrition in order to tackle the overarching challenge of stunting: poor physical and mental development in children. I also kept the problem definitions and underlying causes very short, simple, to the point, and did not embed multiple ideas. My problem statement is a very simple aggregation of the problems, underlying causes, and negative impacts—without the addition of grant proposal-type project introductions.

Simple Project Outline of Problems/Causes/Impacts

Problems

- chronic diarrhea in children;
- chronic under-nutrition.

Underlying Causes

- lack of knowledge of health, hygiene, and family nutrition;
- overall shortage of food and specifically for the four months preceding the corn harvest.

Their negative impacts: Stunting is lack of physical and mental development in children that affects participation in family/community activities. This restricts the children's ability to attend and concentrate in school, leading to a reduction in their ability to develop and prosper as adults. The challenges also reduce the ability of adults to lead the productive, meaningful, prosperous lives they need to leave the cycle of poverty and contribute to the development of their communities.

Problem Statement

Three hundred small children and their 100 families in four Guatemalan villages are frequently ill with chronic diarrhea caused by poor knowledge of health and hygiene, and are chronically under-nourished, caused by little knowledge of nutrition and less than 12 months of family food reserves. These challenges contribute to stunting and restrict the children's ability to attend and concentrate in school, leading to a reduction in their ability to develop and prosper as adults. These challenges also reduce the ability of adults to lead the productive, meaningful, prosperous lives they need to leave the cycle of poverty and contribute to the development of their communities.

Chapter 1 Resources

Suggested Homework Assignment

The complete Field Assignment 1 homework to turn in will be:

1. Write a very short, one-paragraph narrative summary about the community that you worked with in the assessment, about the participants, and about the results.
2. Write out the full list of community-identified needs/problems with the number of votes each received arranged in a simple matrix.
3. A simple project outline of prioritized problems, underlying causes, and negative impacts.
4. A short problem statement that is nothing more than the combination (copy and paste) of the problems, underlying causes, and negative impacts from the simple outline above.

Use the Example of Field Assignment 1 as the template for the assignment.

Course Downloads

Go directly to this book's companion webpage, TimMagee.net/field-guide-to-cba/ to download the following resources.

- Example of Field Assignment 1.
- Participatory needs assessment workshop lesson plan.

Recommended Resources

Website addresses change frequently. Simply enter this book's webpage for current links to resources, or enter the author's name, the organization's name and the document's name into your web browser to find the most current link.

Chatty, D., Baas, S., and Fleig, A. *Participatory Processes towards Co-Management of Natural Resources in Pastoral Areas of the Middle East. Module II: Introducing Participatory Approaches Methods and Tools*, FAO. Available at: ftp://ftp.fao.org/docrep/fao/006/ad424e/ad424e00.pdf

Dayal, R., van Wijk, C., and Mukherjee, N. *Methodology for Participatory Assessments with Communities Institutions and Policy Makers*, World Bank. Available at: http://bscw.ihe.nl/pub/nj_bscw.cgi/d2220629/DayalMethodologyforParticipatoryAssessments.pdf

Jayakaran, R. *Ten Seed Technique*, World Vision International. Available at: http://ravijayakaran.com/books.htm

Theis, J. and Grady, H. *Participatory Rapid Appraisal for Community Development*, IIED. Available at: http://pubs.iied.org/pdfs/8282IIED.pdf

UNDP Bureau of Development Policy. *Designing Climate Change Adaptation Initiatives: A UNDP Toolkit for Practitioners*, UNDP. Available at: http://www.adaptationlearning.net/sites/default/files/17750_CC_un_toolbox_0.pdf

United Nations Development Programme. *Gender, Climate Change and Community-Based Adaptation: A Guidebook for Designing and Implementing Gender-Sensitive Community-Based Adaptation Programmes and Projects*, UNDP. Available at: http://www.undp-adaptation.org/projects/websites/docs/KM/PublicationsResMaterials/Gender_Climate_Change_and_Community_Based_Adaptation_%282%29.pdf

2 Community

Local Climate Knowledge

In Chapter 1, you assessed the development challenges that your community faces. In Chapter 2, you will learn about their understanding of how the climate is changing around them, uncover solutions to climate change challenges which they have developed on their own, and discover where they need capacity building for increased resilience.

In this field assignment, you will begin by facilitating a workshop focused on the climate hazards the community faces—and how a changing climate may impact their livelihoods and increase their vulnerability. The four activities in the workshop will allow you to see the impact of climate change from four different perspectives. Combined, they will allow the community to prioritize the most significant hazards and identify their impacts on livelihood resources and assets. These hazards and negatively affected resources will be used to revise the Chapter 1 project outline.

Chapter 2 will help you accomplish four steps in the development of your project. You will do the following:

1. conduct a community based participatory capacity and vulnerability assessment;
2. summarize and correlate the results of the assessment;
3. identify disaster and climate change challenges, coping strategies and adaptive capacities;
4. revise and update your project outline.

What you'll need:

- access to a community;
- representative community members to participate in the workshop;
- workshop materials, snacks, and drinks.

Timeframe:

- 9 hours:
    - 2 hours to prepare for the workshop;
    - 6-hour workshop;
    - 1 hour for incorporating workshop results into the project outline.

Field Assignment 2 Step 1: Facilitating a Participatory Capacity and Vulnerability Assessment

How can you access local climate knowledge from your community? Community based adaptation to climate change combines local knowledge and scientific knowledge in a way that will empower community members to take charge in an effective bottom-up campaign of adapting to climate change. Their project will be sustainable, as this bottom-up approach gives them project ownership.

Step 1 is to facilitate a workshop which will help in collecting local knowledge about climate-related hazards, the community's vulnerabilities and their adaptive capacities. You will learn from them about the coping strategies they are using in the face of a changing climate.

The goal of this workshop is to determine what the greatest climate hazards are that the community faces, when and where they occur, who is most vulnerable to them, and how they affect livelihood activities, assets and resources. The four activities in the workshop will allow you to see the impact of climate change from four different perspectives. Combined, they will allow the community to prioritize the most important hazards and identify their impacts on livelihood resources and assets.

Organizing a Participatory Capacity and Vulnerability Assessment Workshop

Begin by setting up a 6-hour meeting with 10–12 community members. Meet with community members who represent the ultimate beneficiaries (mothers, fathers, families, farmers, weavers—whoever best describes the community you are working with); try to avoid basing your entire assessment on a meeting with people in powerful positions, for example, the mayors or village leaders. It is important that women and marginalized community members also have a voice in the process. You may need to have several meetings with the different groups so that individuals will feel most comfortable expressing themselves.

You are going to facilitate a 6-hour workshop with four activities. This can also be done in two workshops lasting 3 hours each. The timetable should be as follows:

Seasonal calendar:	1½ hours
Participatory hazard map:	1½ hours
Historical timeline:	1½ hours
Hazard impact matrix:	1½ hours

The activity summary, illustrated handout, and lesson plan for conducting the workshop can be downloaded from Chapter Resources, or can be read in Chapter 10, Field Guide 2. The lesson plan has complete step-by-step instructions for facilitating the workshop.

Review the four activities with your team in advance and make simple adaptations to the activities so they are specific to your community context. You may choose to produce an illustrated poster for the workshop—especially if some participants cannot read. Role-play the activities with your colleagues so that you will be better prepared when you present the workshop, and so you can discover if there are any cultural or linguistic problems.

Activity 1: Seasonal Calendar

Read Activity 1 in Text Box 2.1. Download the lesson plan from Chapter Resources or read it in Field Guide 2, Chapter 10. The goal of this seasonal calendar exercise is to learn more about how the community functions during an annual cycle. From this seasonal calendar, you will learn when the rainy season is, when the dry season is, when farmers do the planting, and when they harvest. You will be able to see when hazards typically occur, such as extreme weather events, heavy rainfall, drought, when diseases may flare up, and when there are hunger gaps in food security.

As you correlate this information, you will begin seeing patterns of how important livelihood activities coincide with seasonal weather patterns which may also coincide with challenges brought upon by climate change.

Box 2.1

Activity 1: Seasonal Calendar

Timeframe: 1½ hours

This exercise is to draw a seasonal calendar in the form of a matrix on a sheet of newsprint—or several sheets taped together. On this calendar you're trying to establish relationships between times of the year, and seasonal events that happen in the community. These could include:

- the rainy season;
- the dry season;
- periods of drought;
- extreme weather events such as flooding;
- important livelihood activities;
- diseases;
- periods of hunger;
- planting and harvesting;
- school term-time;
- annual festivals or ceremonies.

Along the top row of the matrix write the initials for the 12 months of the year. It's helpful to create the matrix the day before the workshop. So that all workshop participants can engage in the activity you can make it very visual by drawing seasonal symbols—such as harvesting maize—so that non-readers will be included.

Along the vertical column on the left you can begin writing down events as community members come up with them. Then, adjacent to the event you can make a mark in the appropriate months that the event occurs. One helpful technique is to have a preliminary piece of paper that you can quickly write down participants' ideas. This will give participants the freedom to speak openly and quickly. After a good number of ideas have been voiced, take a moment to organize the key events since many will be related to each other or simply phrased in a different manner. When you're satisfied with the organization of the events, you can transfer them to the blank calendar.

Table 2.1 Course project example: seasonal calendar

Event	*J*	*F*	*M*	*A*	*M*	*J*	*J*	*A*	*S*	*O*	*N*	*D*
Dry season	X	X	X	X							X	X
Planting				X	X					X		
Rainy season					X	X	X	X	X	X		
Drought								X	X			
Hunger							X	X	X	X		
Harvest			X								X	
Flooding					X	X				X		
School	X	X	X	X	X	X		X	X	X		

Once the calendar has been filled in with events and dates (Table 2.1), introduce the following questions:

- Are the hazards concentrated in one time period or season?
- Are there time periods in the year which are the most difficult for community members and their livelihood assets?

Note the following:

- What are the community members' current coping strategies for dealing with these difficult periods?
- Capacity building: Which of the difficult periods are they having trouble coping with due to a lack of strategies?

Activity 2: Participatory Climate Hazard Map

Read Activity 2 in Text Box 2.2. Download the lesson plan from Chapter Resources or read it in Field Guide 2, Chapter 10. The goal of this hazard mapping exercise is to learn more about spatial relationships and areas at risk within the community. You want to look at the community's geographical layout including the location of the village proper in relationship to farm fields, hills, forests, roads, and water sources. You also want to look at infrastructure such as houses, social areas, schools, and community resources such as storage areas—and safe havens where community members can go in case of emergency.

Next is to begin the process of overlaying hazards onto the map. These hazards might be floods, sections of the community that are most affected by drought or by heavy rain. The information that you want from the exercise is to identify which parts of the community, which people within the community, which assets, which environmental resources, and which livelihoods are the most vulnerable to the hazards as identified on the map.

Activity 3: Historical Timeline

Read Activity 3 in Text Box 2.3. Download the lesson plan from Chapter Resources or read it in Chapter 10, Field Guide 2. The aim of this historical timeline exercise is to learn more about past and current hazards, their frequency, and if they are becoming more frequent or

Box 2.2

Activity 2: Participatory Climate Hazard Map

Timeframe: 1½ hours

This exercise involves drawing a participatory hazard map of the community. Participatory mapping is an inclusive tool because all workshop participants can engage in the activity as it is very visual—non-readers will be included.

Consider returning to the village the day before this exercise to tour the farm fields, forests, and water sources with one of the villagers. Take a few minutes to talk to people you meet in order to gain a greater understanding of some of the challenges they are facing.

Draw the community map on a sheet of newsprint—or several sheets taped together. Begin by making a very simple drawing of the spatial relationships between the different parts of the community and how the village relates to the farm fields, hills, very steep hills, and sources of water. Mark where transportation routes are, including roads and pathways. One suggestion is to quickly draw a 10-minute preliminary map, make corrections and adjustments, and then transfer the revised map information to a fresh sheet of paper for further development.

When everybody at the workshop is satisfied that the basic map represents the community, farming areas and surrounding environmental resources, you can begin marking things on the map such as where individual homes are and where their farm fields are. It is a good idea to locate buildings and farmers' plots using pieces of colored paper that can be attached to the map with removable tape so they can be moved or adjusted. The paper cutouts are also useful because they can be completely removed if you want to go back to the basic map for a future workshop on a different issue.

When everyone is satisfied that the map is accurate, introduce the idea of hazards that the community suffers. These hazards could include extreme weather events, floods, heavy rainfall, drought and landslides. Look back at Activity 1 for additional ideas.

Once the hazards have been indicated on the map, introduce the following questions:

- Are the hazards concentrated in one area of the community?
- What negative impacts will the hazards have on community members and their assets and resources?
- Who in the community is the most at risk from the hazards?
- Are there safe places in the neighborhood where community members can shelter from the hazards?

Note the following:

- What are the community members' current coping strategies for dealing with these difficult events?
- Capacity building: Which of the difficult events are they having trouble coping with due to a lack of strategies?

To see what this could look like, go to Chapter 10, Field Guide 2.

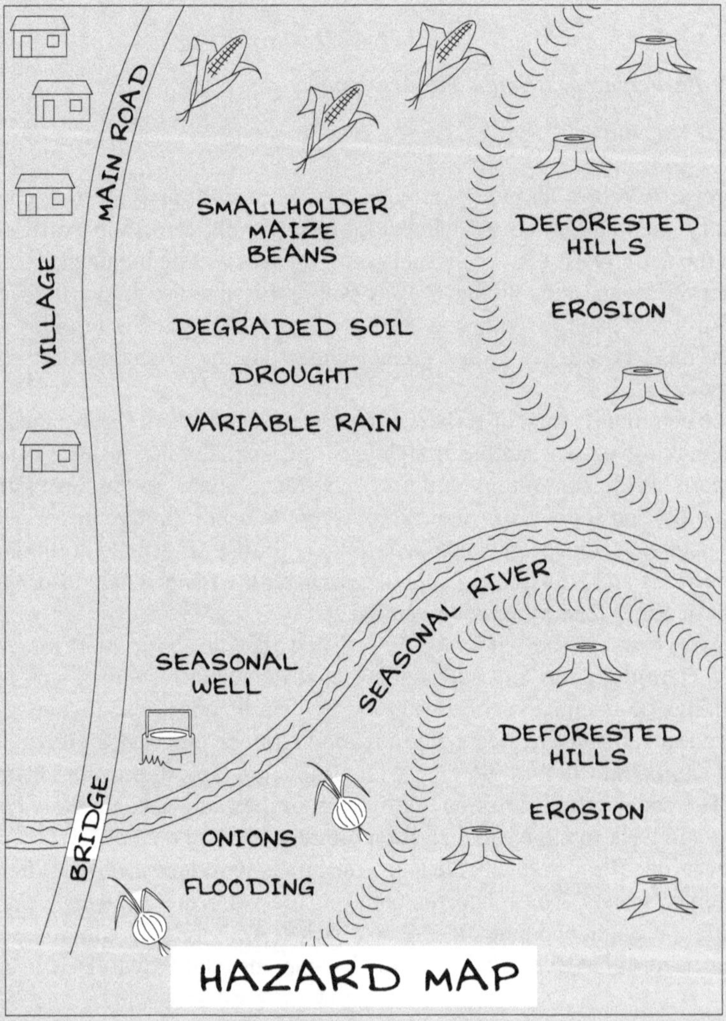

more intense with time. The historical timeline can also pinpoint changes in livelihoods, crops, and in land use. It can also reveal if and when water resources became scarcer, and if and when farm lands became less productive.

It's best if this exercise can be carried out in cooperation with older members of the community who have a long-term view of events. In Chapter 3 you will compare the changes in weather and extreme weather events identified by the community with actual meteorological data.

Activity 4: Climate Hazard Impacts on Livelihoods

Read Activity 4 in Text Box 2.4. Download the lesson plan from Chapter Resources or read it in Chapter 10, Field Guide 2. The aim of this exercise is to determine the impact of the

Box 2.3

Activity 3: Historical Timeline

Timeframe: 1½ hours

The historical timeline is an exercise that is a very simple matrix with years in the left column and important events in the right column (Table 2.2). You will be looking for insights into past hazards and events, and how they may have changed or intensified over time.

Table 2.2 Course project example: historical timeline

Year	*Hazards and events*
2011	10-day rainstorm in October drops up to 5 feet of rainwater in parts of Guatemala
2010	Hurricane Agatha
2009	Recurring drought in August and September
2008	Noticeable shift in the beginning and end of the rainy season; unpredictable rainfall
2005	Hurricane Stan
2004	Famine in the Polochic region of Guatemala
2001	Government relocates returning refugees adjacent to the village
2000	Hurricane Gordon
1996	End of the Civil War
1995	Hurricane Mitch
1976	7.6 magnitude earthquake

These could include hurricanes, droughts, health challenges, famines or floods. Hopefully, there will be some village elders participating in the workshop and that will allow you to get a long-term perspective from 20 or 25 years ago so that you and the participants can see if these events are occurring more frequently.

Next, when the group has completed the timeline, introduce the subject of climate change. Have they seen a change over time with climate change challenges? When did they notice the changes? Some examples:

- Beginning 20 years ago, rainfall began decreasing; by how much?
- Beginning 20 years ago, the growing season changed; it's shorter now—or it starts later.
- Beginning 20 years ago, storms have increased and there is flooding now when there did not used to be flooding.
- Beginning 20 years ago, we've had to walk progressively further to get water. How much further?

Note the changes which the participants have seen and briefly describe how they've changed and over what time frame. Does the community realize this is linked to climate change and realize that this may be ongoing?

Note the following:

- What are the community members' current coping strategies for dealing with these difficult events?
- Capacity building: Which of the difficult events are they having trouble coping with due to a lack of strategies?

hazards you have identified on livelihood assets and resources, and to determine where community members are the most vulnerable. Take a few minutes alone to list both the hazards, and the livelihood assets and resources from the first three activities. This is simply to assist in starting a discussion in order to begin filling in the matrix.

Box 2.4

Activity 4: Climate Hazard Impacts on Livelihoods

Timeframe: 1½ hours

This matrix is another very simple matrix with important livelihood resources and assets in the left column and important hazards in the top row. It's a good idea to prepare the blank matrix on newsprint in advance. Also, take a few minutes alone to list both the hazards and the livelihood assets and resources from the first three activities onto a sheet of notebook paper. These can be used to start a discussion in order to begin filling in the matrix. Doing a quick preliminary matrix on a blank sheet of newsprint during this discussion is also a good idea—you can then just transfer the list of assets and hazards onto a clean, blank matrix.

Typical hazards may include:

- extreme weather events, such as hurricanes or cyclones;
- flooding;
- drought/heat waves or cold fronts/heavy snow;
- unpredictable beginning and end to the rainy season;
- erratic rainfall or more or less rainfall;
- change in the timing of the growing season;
- health issues/diseases;
- insect infestations;
- crop and livestock disease outbreaks.

Important assets and resources may include:

- income generation from agriculture;
- cooperatives, associations, management committees;
- crop land;

- crop productivity;
- livestock;
- irrigation systems;
- health;
- food reserves/food security;
- environmental resources such as forests and water.

Table 2.3 Course project example: climate hazard impacts on livelihoods

Shaded area indicates most critical hazard/livelihood impacts prioritized by the community

Livelihood	*Change in growing season*	*Midseason drought*	*Erratic rain*	*Extreme tropical storms*
Food security and nutrition	3	2	3	1
Income generation	3	3	3	2
Crop Productivity	3	3	3	2
Access to water	0	3	3	1

Field Assignment 2 Step 2: Summarizing and Correlating the Results of the Assessment

Activity 1: Summarize the Key Points

In this activity you are simply going to summarize the key points from the workshop's four activities. Go to Text Box 2.5 to see what this could look like.

Box 2.5

Course Project Example

Seasonal Calendar

Prioritize and list the events which occur during the year. Briefly describe how they've changed. Does the community realize that some are linked to climate change and realize that this may be ongoing? Here are the results from the Example of Field Assignment 2.

The period from May through October appears to be the most critical due to the number of important events and to the fact that it is the rainy season.

Traditional events and challenges are:

- planting and harvesting of crops;
- rainy season for growing subsistence crops;
- shortage of food for the four months preceding the harvest.

Climate change-related events:

- unpredictable beginning and end to rainy season;
- erratic rainfall during rainy season;
- intermittent drought during rainy season;
- extreme weather events and flooding.

The five-month rainy season is very important for the community since it allows them to grow their non-irrigated subsistence crops such as maize and beans. Having an unpredictable start date for the rainy season makes planning for planting difficult. Mid-season drought can impact crop development, and an early end to the rainy season can mean that crops do not reach full maturity. Extreme weather events such as tropical storms and hurricanes can cause flooding, destroy crops, erode topsoil, and damage livelihood assets.

These challenges have a negative impact on food security and nutrition, levels of crop production, income generation, livelihood assets, and access to water. The workshop participants did not realize that these negative events may be related to a changing climate.

Climate Hazard Map

Note the hazards which the participants indicated on the hazard map. Briefly describe where the hazards are concentrated and which community members are the most affected. Does the community realize that some are linked to climate change and realize that this may be ongoing? Here are results from the Example of Field Assignment 2.

The main hazards that the community voiced during this mapping exercise were:

- flooding of low-lying fields during heavy rain and tropical storms;
- drought;
- erosion of topsoil and gully formation on steeply sloping fields.

These hazards create the greatest challenges for farmers and their fields. These challenges reduce crop productivity, affect food security and nutrition for their children, access to water, and family income. The workshop participants did not realize that these negative events might be related to a changing climate.

Historical Timeline

Is the community seeing a change in frequency or intensification of hazards and challenges? Please note the changes which they've seen. Briefly describe how they've changed and over what timeframe. Does the community realize that some are linked to climate change and realize that this may be ongoing? Here are results from the Example of Field Assignment 2.

This prioritized list of climate-related challenges and hazards shows what the participants identified as becoming more frequent or having intensified over the past 25 years:

- change in the timing of the growing (rainy) season; it seems to be starting later and ending earlier;
- August traditionally had a two-week dry period during the rainy season—the dry period is becoming longer;
- erratic rainfall during the rainy season; rainfall has become increasingly erratic;
- increase in frequency of tropical storms and extreme weather events (now every two to three years);
- flooding.

The increase in the frequency of extreme weather events began more than 20 years ago in 1995 with Hurricane Mitch. Hurricanes and tropical storms happen approximately every two to three years, are quite extreme, and cause damage to crops and farm fields. Although the community is aware that their increased frequency began in recent history, they do not correlate this to a changing climate.

Prioritization of Climate Hazard Impacts on Livelihoods

What is the prioritization of the community's greatest hazards they face? Which parts of their lives are the most vulnerable? Summarize the hazards and the livelihood assets and outcomes from the Activity 4 as prioritized by the community. Here is the summary from the Example of Field Assignment 2.

The greatest hazards that the community voiced during this exercise were prioritized as:

- unpredictable beginning and end to rainy season;
- intermittent drought during rainy season;
- erratic rainfall during the rainy season;
- increase in frequency of tropical storms and extreme weather events (now every two to three years);
- flooding.

These climate hazard challenges have a direct and negative impact on these prioritized livelihood assets and resources:

- food security and nutrition;
- income generation;
- crop productivity;
- access to water.

Adding information from Activities 1 and 2, the most critical time period for these hazards and the livelihoods they impact is the period from May through October. The community members most severely affected are farmers and their families.

Field Assignment 2 Step 3: Coping Strategies and Adaptive Capacity

Activity 1: Coping Strategies

During the four activities you asked the participants about traditional coping techniques they have used to address the challenges that they identified. Summarize the techniques which they mentioned.

Water seems to be a recurring challenge for the community based upon the results of the workshop. The community was asked at the end of each of the four activities if they had found strategies for coping with some of these challenges. It does not appear that they had successful coping strategies. They have little knowledge of water harvesting, conservation or management and little knowledge of drought-resistant cropping systems.

Activity 2: Adaptive Capacity

The adaptive capacity of the community members can be measured using variables of the five capitals: human, social, physical, natural, and financial (Table 2.4). Our understanding of the community's strengths and weaknesses in these five capitals will help us improve the capacity-building components of our project outline.

Table 2.4 The five capitals: examples of possible livelihood assets and resources

Capital	*Definition*
Human capital	Understanding of future change, knowledge about adaptation options, ability to assess them, and capacity to implement them
Social capital	Community management committees (water, agricultural, forest)
Physical capital	Infrastructure and equipment and improvements in crops and livestock resources
Natural capital	Productivity of land, water, and biological resources for which livelihoods are derived
Financial capital	Alternative sources of income generation; market access for products and services; micro-savings programs

Box 2.6

Course Project Example

Human capital

The community has very little knowledge about climate change and how it may impact their lives. Similarly, they are unaware of strategies that they can use to adapt to climate change.

- *Possible project solution*: An introductory, consciousness-raising workshop about climate change, its impacts, and activities that they can incorporate into their lives in order to increase their resilience.

Social capital

Currently, the community has no oversight committees to help manage their livelihood assets and resources.

- *Possible project solution*: Facilitate the development of a project committee with subcommittees for developing a water use management plan, a land use management plan, and a farmers association.

Physical capital

The community has no infrastructure to speak of.

- *Possible project solution*: A community water management committee could investigate alternate sources of water for when it's dry—and diversion canals or terraces for controlling water during periods of heavy rain.

Natural capital

What were once lush forests have been over-harvested for firewood and lumber. This complicates water management because without the trees, rain simply runs quickly off of the hills and isn't able to infiltrate into the underground natural water system. Runoff erodes soil and contributes to flooding.

- *Possible project solution*: A reforestation plan, and workshops on soil conservation and on water conservation and management for the farmers.

Financial capital

These are largely subsistence farmers. Although a few of them do sell onions, they are not well organized and do not have consistent market contacts.

- *Possible project solution:* Facilitate the formation of a farmers association to survey local and regional businesses which purchase agricultural produce and to determine what products they need on a routine basis. This could give them a market link that they need and possible training for fulfilling market demand.

Save these possible solutions for Chapter 5 when you begin adding solution-oriented activities to your project outline.

Field Assignment 2 Step 4: Revising the Project Outline Based upon New Findings

What is interesting about the series of activities in this chapter is that you've been able to assess community need through several different lenses.

In Field Assignment 1, you facilitated an open-ended discussion with the community to find out what their needs, wants, and challenges were. From there you organized the

Box 2.7

Course Project Example

Table 2.5 Comparing and combining needs and challenges from Field Assignments 1 and 2

Outline from Field Assignment 1	*Livelihood assets and hazards from Field Assignment 2*
Problems:	Livelihood assets and resources negatively impacted by hazards:
Chronic diarrhea in children	Food security and nutrition
Chronic under-nutrition	Income from agriculture
	Crop productivity
	Access to water
Causes:	Hazards:
Lack of knowledge of health, hygiene, and nutrition	Unpredictable beginning and end to rainy season
Overall shortage of food and specifically for the four months preceding the corn harvest	Intermittent drought during rainy season
	Erratic rainfall during the rainy season
	Increase in frequency of tropical storms and extreme weather events (every two to three years)
	flooding

By comparing the two assessments side-by-side you can see that there are similarities with food security and nutrition. The vulnerability and capacity assessment also added a new problem: insufficient agricultural income. This increases their vulnerability in times of stress and is directly related to the newly identified underlying causes (hazards in the list above). Let's add insufficient agricultural income to the problem section of the project outline.

The community in the original needs assessment did not voice any climate change-linked causes. In the vulnerability and capacity assessment they raised the fact that a change in the timing of the growing season, drought, unpredictable rainfall during the rainy season, and extreme tropical storms have reduced the productivity of their crops and access to water. These affect food security and nutrition, and reduced agricultural income. Let's add these climate change-linked challenges to the underlying causes section of the project outline.

The climate hazards identified by the community are clues to ideas that we can use in developing activities for the project in Chapter 5. For example, a change in the timing of the growing season could lead us to consider an activity that would explore drought-resistant or early maturing crops. Erratic rainfall during the rainy season could lead us to consider an activity that could include soil restoration by increasing organic material to better hold water and by adding mulch to reduce evaporation. You could also consider water harvesting and water conservation activities.

information into a simple project outline with one or two over-arching problems and their underlying causes. It is possible that the majority of the needs and underlying causes from that exercise had a basis in traditional development—not in climate change-linked challenges.

In Field Assignment 2, you have taken an approach which allowed you to identify climate change-linked disasters and hazards from several different perspectives and then correlate all of these with impacts on livelihoods and assets.

How do the challenges identified in this relate to your project's original Chapter 1 outline? Based upon the results of the workshop, would you make any modifications to your original project?

- Do you need to add a problem that is a priority which was not included in your original outline?
- Do you need to modify the definition of one of your underlying causes now that you have more information?
- Does the community have coping strategies that would be good to include in your outline?

You have a new list of the top hazards and livelihoods from the summary in Activity 4 of Step 3. Compare this new list to the simple project outline from Field Assignment 1.

Write a short paragraph about any modifications that you may need to make to the project outline to reflect this newly discovered local climate knowledge.

Finally, make the modifications to the project outline and to the problem statement. Remember to keep the problem statement absolutely parallel to the outline.

You can see how this develops in the Example of Field Assignment 2 in Box 2.7.

Revised Project Outline

Let's look at the problems/causes/impacts and incorporate new information from the vulnerability and capacity assessment.

Problems

- Chronic diarrhea in children.
- Chronic under-nutrition.

New problem identified in Field Assignment 2 activities:

- Insufficient income from agriculture.

Causes

- Lack of knowledge of health, hygiene, and family nutrition.
- Overall shortage of food and specifically for the four months preceding the corn harvest.

New underlying cause related to climate change:

- Unpredictable dates for the start and end of the rainy season, intermittent drought and erratic rainfall during the rainy season, flooding and extreme weather events have reduced crop harvests and access to water.

Revised Problem Statement

Three hundred small children from 100 families in four Guatemalan villages are frequently ill with chronic diarrhea caused by little knowledge of health and hygiene, are chronically under-nourished caused by little knowledge of nutrition and less than 12 months of food reserves. Their families suffer from insufficient agricultural income from reduced crop harvests and access to water due to unpredictable dates for the start and end of the rainy season, intermittent drought and erratic rainfall during the rainy season, flooding, and extreme weather events. These challenges contribute to stunting and restrict the children's ability to attend and concentrate in school, leading to a reduction in their ability to develop and prosper as adults. These challenges also reduce the ability of adults to lead the productive, meaningful, prosperous lives they need to leave the cycle of poverty and contribute to the development of their communities.

Conclusion

It is interesting to observe that the hazards and the livelihood assets and resources were similar and consistent between the four workshop activities. On top of that, several of them are underlying causes to the problems which the community expressed in Chapter 1. For example, the climate challenges which have led to reduced crop productivity are an important underlying cause to the challenge of chronic under-nutrition.

Chapter 2 Resources

Suggested Homework Assignment

The complete Field Assignment 2 homework to turn in will be:

1 a hazard impact matrix showing:

 a a prioritized list of hazards identified by the community on the top row;
 b a prioritized list of livelihood assets and resources on the left column;
 c the results of a vote by the community on the matrix prioritizing which hazards make which livelihood assets and resources the most vulnerable.

2 a short bullet-point summary of the hazards and livelihoods results of each of the four activities;
3 a simple comparison between your project outline from Field Assignment 1 and Activity 4, Step 3 summary of assets and hazards;
4 a brief description of the changes you might make to your project outline;
5 a revised project outline including one or two new problems or underlying causes identified in Field Assignment 2.

Use the Example of Field Assignment 2 as the template for the assignment.

Course Downloads

Go directly to this book's companion webpage, TimMagee.net/field-guide-to-cba/ to download the following resources:

- Example of Field Assignment 2;
- Participatory Capacity and Vulnerability Assessment Workshop Overview and How-to Card;
- Participatory Capacity and Vulnerability Assessment Workshop Lesson Plan:
  - Seasonal calendar activity;
  - Hazard mapping activity;
  - Historical timeline activity;
  - Climate hazard impacts on livelihoods.

Recommended Resources

Website addresses change frequently. Simply enter this book's webpage for current links to resources, or enter the author's name, the organization's name and the document's name into your web browser to find the most current link.

CARE. *Climate Vulnerability and Capacity Analysis Handbook*, CARE. Available at: http://www.careclimatechange.org/files/adaptation/CARE_CVCAHandbook.pdf

Carloni, S. *Rapid Guide for Missions: Analyzing Local Institutions and Livelihoods*, FAO. Available at: ftp://ftp.fao.org/docrep/fao/008/a0273e/a0273e00.pdf

Chatty, D., Baas, S., and Fleig, A. *Participatory Processes towards Co-Management of Natural Resources in Pastoral Areas of the Middle East. Module II: Introducing Participatory Approaches Methods and Tools*. FAO. Available at: ftp://ftp.fao.org/docrep/fao/006/ad424e/ad424e00.pdf

Dayal, R., van Wijk, C., and Mukherjee, N. *Methodology for Participatory Assessments with Communities Institutions and Policy Makers*, World Bank. Available at: http://bscw.ihe.nl/pub/nj_bscw.cgi/d2220629/DayalMethodologyforParticipatoryAssessments.pdf

Krantz, L. *The Sustainable Livelihood Approach to Poverty Reduction*, SIDA. Available at: http://www.sida.se/Svenska/Om-oss/Publikationsdatabas/Publikationer/2003/september/The-Sustainable-Livelihood-Approach-to-Poverty-Reduction/

Regmi, B., Morcrette, A., Paudyal, A., Bastakoti, R., and Pradhan, S. *Participatory Tools and Techniques for Assessing Climate Change Impacts and Exploring Adaptation Options, Livelihoods and Forestry Programme*. Available at: http://www.lfp.org.np/publications.php?id=34

Theis, J. and Grady, H. *Participatory Rapid Appraisal for Community Development*, IIED. Available at: http://pubs.iied.org/pdfs/8282IIED.pdf

3 Climate

Scientific Climate Information

In this chapter you will analyze scientific and meteorological information specific to your community's location and determine what are the climate change scenarios happening now and which scenarios will likely play out in the future.

Where Can You Find Scientific Information about Your Community's Specific Location?

You will be using online resources to collect information, and consulting with a local expert; this could be a government official or specialist at an NGO that has worked with climate change projects. You're going to be looking for both the current trends of a changing climate, as well as an overview of potential future impacts. You will do the following:

1 conduct an online literature review;
2 investigate climate change scientific studies to collect information for your location;
3 consult with a local expert;
4 compile the information.

The goal will be to gather data on local climate stimuli such as changes in temperature, precipitation, surface water flow, and climate variability. The sources you discover will provide you with projected impacts, such as changes in agricultural production, changes in water availability, impacts on livelihood, and impacts on biodiversity and ecosystems. You will compile this data into a table to compare it to your community's experience with a changing climate identified in Chapter 2.

This comparison can be very important in your plan for project sustainability and community ownership. If you discover scientifically-backed information that compares favorably with local climate knowledge, then you can begin building bridges between scientific knowledge and the community's local knowledge. If your community has had problems finding coping strategies—or if their coping strategies are not as effective as they could be—project activities that have shown scientific evidence of having worked can now be introduced to them across this bridge. It is a very powerful strategy to offer potential improvement to coping strategies that the community has developed for themselves.

Your project outline from Chapter 2 likely had both traditional development components and adaptation components. The data that you collect in this assignment will help you to determine with greater certainty if your CBA project components are indeed related to a changing climate.

Why is It Important to Know if CBA Components are Linked to Climate Change?

A single, unique development activity may be used to address traditional community development challenges, disaster risk reduction, or adaptation to climate change challenges. For example, reforesting steep hillsides behind a community's village could be considered a development project if it will help recharge the village's spring that has dried up due to deforestation, or it could be considered a disaster risk reduction project if it will help prevent flooding in the village by reducing runoff.

But if the community's spring ran dry because of a long-term, climate change-linked drought, then the reforesting project may not be the most appropriate solution to solve that specific problem. Knowing this will allow you to focus scarce resources on activities that will help community members adapt to—in this example—a shortage of water and a long-term drought.

Chapter 3 will help you accomplish three steps in the development of your project. You will do the following:

- research scientific evidence of climate change for your community;
- identify current and projected climate change impacts;
- summarize information into a table for comparison with local climate knowledge.

What you'll need:

- access to an Internet browser or university search engine;
- a local climate change expert.

Timeframe:

- 3 hours:
    - 1 hour of online research;
    - 1 hour for meeting with climate change expert;
    - 1 hour for information compilation.

Field Assignment 3 Step 1: Researching Scientific Climate Change Information

In this assignment you will be conducting an online literature review to determine which climate-related stimuli are impacting your community, what the projected change in those stimuli will be, and what their projected negative impacts might be on the community.

What are Climate-Related Stimuli?

Climate change-related stimuli typically include changes in temperature and moisture, but can also include variability from the norm in temperature and moisture. Typical climate-related stimuli could include:

- temperature;
- precipitation;
- surface water;
- changing rainfall patterns.

Activity 1: Online Search Terms: Defining the Climate Change Problem

Clearly define your location and your community's challenges by sector (health, food security, agriculture, disaster risk reduction). The definition needs to be one short line as you are going to be using it as a search term in Internet browsers. An example for the project could be: "Guatemala, climate change impacts, changing temperature and precipitation, food security, and agriculture." These terms can be arranged and rearranged in an effort to optimize search results. Here are two examples:

- Guatemala + climate change impacts on agriculture;
- Are temperatures increasing in Guatemala due to climate change?

Depending on the theme of your project, you will most likely be looking for changes in temperature, precipitation, and water availability—and to determine what those changes are, whether they have already begun, whether they are anticipated in the future, and what their negative impacts will be.

Where's the Evidence for Climate Change?

Conduct an online review for information. This is an informal review—but one that can generate good search results due to the specificity of your community's definition. Make a note of the titles and links for the information that you find and write a short paragraph summarizing their content.

You will be rewarded with a wide range of information. Try and avoid the popular press and focus instead on studies performed by respected institutions. Using your Internet browser—or preferably a university search engine—search for peer-reviewed scientific papers, or systematic reviews that address your community's location and challenges they face. Make a note of the titles and links to the scientific papers and write a short paragraph summarizing their content. Go to Box 3.1 to see what this could look like.

Activity 2: Meet with a Local Expert

Find someone who has had experience in climate change or in adaptation in your community's location whom you could meet, speak to on the telephone, or correspond with by e-mail. This person could be from a governmental weather office, or they could be a climate change specialist at an NGO.

Your goal in doing this is twofold. First, you want to see if they have information that corroborates or adds to what you learned in your literature search. Second, it could be very useful for you to be able to connect with an expert whom you can contact again as you continue to more fully develop your project.

Go to Box 3.1, Example of Field Assignment 3, to see what this could look like.

Box 3.1

Course Project Example

Field Assignment 3 Step 1: Researching Scientific Climate Change Information

Part 1 of this step is defining the climate change problem: assessing local climate change information online. I defined my areas of interest as Guatemala, climate change (CC) impacts, changing temperature and precipitation, food security, and agriculture. I used various combinations of these terms in my Internet search engine to search for literature. There were quite a few papers discussing a range of CC challenges, but I focused only on those papers that specifically addressed my areas of interest above.

My question on the Internet was: "Are temperatures increasing in Guatemala due to climate change?" The following web addresses turned up and are summarized below.

Adaptation Fund (2010) Proposal for Guatemala. Online. Available at: http://adaptation-fund.org/system/files/AFB.PPRC_.2.6%20Proposal%20for%20Guatemala.pdf (accessed 12 April 2012).

Guatemala faces many hazards related to climate variability and climate change. Projections show increases in temperature, decreases in total mean precipitation, increases in the frequency of extreme precipitation events, as well as in the frequency and intensity of extreme climatic events. Some 10 percent of Guatemalan territory faces risk of drought and more than 3,000 communities are prone to flooding.

Lennox, J. (ed.) (2010) *The Economics of Climate Change in Central America: Summary 2010*. Available at: http://www.eclac.org/cgi-bin/getProd.asp?xml=/publicaciones/xml/9/41809/P41809.xml&xsl=/mexico/tpl-i/p9f.xsl&base=/mexico/tpl/top-bottom.xslt (accessed 12 April 2012).

This study indicates that climate change could cost Central America at least half its gross domestic product by the end of this century as more extreme weather, lower crop yields and water shortages are expected.

The report warns that agriculture will be one of the hardest-hit sectors: "Climate change could significantly affect food security by reducing food production and curbing direct access to food among rural families, as well as leading to higher food prices."

The study predicts that water shortages will also cause falling production of the region's staple foods: maize, rice, beans, and coffee—a key export. These crops are particularly sensitive to rising temperatures and declining rainfall levels, the report says. An average reduction in rainfall levels for Central America of 10–28 percent is expected. In Guatemala, for example, the report forecasts that a 3.5 degree Celsius rise in temperature together with a 30 percent reduction in rainfall could lead to a 34 percent decline in maize production and 66 percent for beans.

World Bank (2009) *Guatemala Country Note: Climate Change Aspects in Agriculture*. Online: Available at: http://siteresources.worldbank.org/INTLAC/Resources/Climate_GuatemalaWeb.pdf (accessed 12 April 2012).

This study expects a median temperature increase of 3.0°C by the year 2050 and a reduction in precipitation during the normal rainy season, periods of drought and an intensification of heat waves with serious implications for agriculture. It is expected that evapo-transpiration will increase due to temperature increases and precipitation reduction, resulting in the expansion of semi-arid areas. The study indicates that there could be decreases in the production of crops of up to 34 percent for corn and up to 66 percent for beans. It expects a reduction of superficial water flow of between 10 percent and 50 percent, which, along with the predicted temperature increases and precipitation reduction, will result in a lower water supply for agricultural irrigation.

Field Assignment 3 Step 2: Assessing Risk and Summarizing Local Scientific Climate Change Information

Activity 1: Compile the Information

Make a very simple table with four columns and an initial five rows. Add these headings to the top of the four columns:

- climate change stimuli;
- projections of changes in stimuli;
- projected negative impacts;
- source of information.

Add each of these stimuli individually the cells in the left column (based upon your research, your stimuli may differ).

- temperature;
- precipitation;
- surface water;
- changing rainfall patterns.

Paste summary information from your online research and from your expert interview into the appropriate cells on the table. Much of the information may be repetitive and so it should be condensed. The goal here is a concise, short document. If sources differ in their increase in temperature, for example, then show a range for the projection from low to high.

Go to Box 3.2, Example of Field Assignment 3, to see what this could look like.

Box 3.2

Course Project Example

Activity 1 Compile the information

Table 3.1 Course project example: climate-related stimuli, changes in stimuli, and projected impacts

Climate change stimuli	*Projections of changes stimuli*	*Projected negative impacts*	*Source of information*
Temperature	Median temperature increase of 3.0°C by the year 2050 and an increase in heat waves	Serious implications for agriculture, an expansion of semi-arid areas; a reduction in food production leading to a reduction in food security; a curbing of direct access to food among rural families; higher food prices	World Bank (2009) Lennox (2010) ; United Nations ECLAC Report
Precipitation	Decrease in precipitation of 9 percent and an increase in drought periods by 2050 Increases in the frequency of extreme precipitation events, as well as in the frequency and intensity of extreme climatic events	Falling production of the region's staple foods including maize, rice, beans, and coffee—Guatemala's chief export; decreases in the production of maize of up to 34 percent; decreases in the production of beans of up to 66 percent	World Bank (2009); Lennox (2010); United Nations ECLAC Report The Adaptation Fund (2010)
Superficial water flow (surface water)	Reduction of between 10 and 50 percent	Lower water supply for irrigation in agriculture	World Bank (2009)
Changing rain patterns	Unpredictable start of rainy season; periods of drought in the middle of the rainy season	Subsistence crops providing declining yields leading to widespread malnutrition and increasing poverty	Tim Magee Climate Change Consultant Guatemala City

Source: Adapted from Kropp and Scholz (2009).

Chapter 3 Resources

Suggested Homework Assignment

The complete Field Assignment 3 homework to turn in will be:

1 Context: Clearly define your location and your sectoral areas of interest (key words for your search).

2 Using a search engine, search the Internet for climate change documents specific to your community's context. Please provide the titles and links to the papers with a short paragraph summarizing their content.
3 Using a university search engine, or your own Internet browser, search for scientific papers that address your community's climate change context. Please provide the titles and links to the papers with a short paragraph summarizing their content.
4 Contact a local climate change expert and discuss your community's climate change context. Provide this expert's name and a summary of your discussion.
5 Develop a chart as found in Example of Field Assignment 3, detailing stimuli observations and impacts that relate to your climate change element in your project. Provide short summaries of the studies that you read and the person you spoke with and provide the source information.

Use the Example of Field Assignment 3 as the template for the assignment.

Course Downloads

Go directly to this book's webpage, TimMagee.net/field-guide-to-cba/ to download the following resources.

- Example of Field Assignment 3.

Recommended Resources

Website addresses change frequently. Simply enter this book's webpage for current links to resources, or enter the author's name, the organization's name and the document's name into your web browser to find the most current link.

CARE. *Community-Based Adaptation Toolkit*, CARE. Available at: http://www.careclimatechange.org/files/toolkit/CARE_CBA_Toolkit.pdf

World Bank. World Bank Climate Change Portal. Available at: http://sdwebx.worldbank.org/climateportal/index.cfm

4 Challenge

Local Context

Comparing Local Climate Knowledge to Scientific Climate Knowledge

In Chapter 4 you will compare your community's local climate knowledge with the scientific climate knowledge from Chapter 3 and verify that the community's CBA challenges are linked to climate change. You will write a context statement that accurately describes your community's specific, local, climate change context. Lastly, you will revise your project outline with the new findings.

Using your scientific research from Chapter 3, you will have the opportunity of looking through a new lens at your community's traditional development challenges and at the CBA challenges developed in Field Assignments 1 and 2.

One of four things can result from this:

1. Your original project outline will do a good job of addressing the community's climate change challenges based on the scientific input from Field Assignment 3.
2. The outline will do a good job of addressing these challenges with minor modifications.
3. You will realize you need to add an additional problem or underlying cause to your outline.
4. On the other hand, you might discover that you were off by a mile with your early CBA challenges. You might have developed your first underlying CBA cause about flooding—when indeed you learned in the scientific research that flooding is not a challenge that you need to be concerned about—but that drought is.

This chapter is your chance to perfect the CBA element of your original project with corrections, modifications, or additions. Chapter 4 will help you accomplish three steps in the development of your project. You will:

1. compare local climate knowledge and scientific climate knowledge;
2. summarize current and projected climate change impacts;
3. revise the project outline to reflect local need, and local and scientific climate knowledge.

What you'll need:

- access to a computer.

Timeframe:

- 2 hours:
  - 30 minutes to summarize impact projections;
  - 30 minutes to compare local and scientific knowledge;
  - 30 minutes for the community climate change context;
  - 30 minutes for the problem statement and project outline modifications.

Field Assignment 4 Step 1: Comparing Local and Scientific Climate Knowledge

Are Your Project Challenges Linked to a Changing Climate?

In this first step you will be comparing your project's challenges to the challenges that local scientific evidence is revealing.

Activity 1: What are the Typical Near-Term and Long-Term CC Challenges that Your Research Revealed?

1 Compile a short paragraph of the most important near-term climate change challenges from your Field Assignment 3 climate change-related stimuli and projected impacts chart.
2 Compile a short paragraph on the most important long-term climate change challenges from your Field Assignment 3 climate change-related stimuli and projected impacts chart.

Go to Text Box 4.1 to see what this could look like.

Box 4.1

Course Project Example

Activity 1: What are the Typical Near-Term and Long-Term CC Challenges that Your Research Revealed?

I reviewed the climate-related stimuli and projected impacts chart from Field Assignment 3 and summarized the information below.

Summary of Near-Term CC Challenges

The near-term climate change challenges in Guatemala are an increasing number of torrential tropical storms and the fact that the dates for the rainy season—which traditionally began in mid-May and ended in mid-October—are no longer as predictable. This makes it difficult for farmers to decide when to plant. The seasonal rains are also increasingly variable, and there are dry periods in the normal rainy season which affect the level of crop production. These challenges reduce water supplies, lead to declining quality and quantity of crop yields, reduce family incomes and food security, and increase food prices and chronic malnutrition in rural areas.

The near-term adverse climate change-related stimuli are:

- unpredictable dates for the beginning and end of the rainy season;
- decreasing precipitation;
- extended dry periods during the rainy season;
- an increasing number of increasing precipitation events and extreme climatic events.

Summary of Long-Term CC Challenges

My research from Chapter 3 revealed long-term climate change challenges over the next 40 years. Scientists are expecting a median temperature increase of 3.0°C, a decrease in precipitation of 9 percent, and the reduction of surface water of between 10 and 50 percent. These complications will lead to an expansion of semi-arid areas, a further reduction in food production, higher food prices, and less water for agricultural irrigation—contributing to an even greater reduction in food security, and an increase in malnutrition and extreme poverty.

The long-term projected climate change related stimuli are:

- median temperature increase of 3°C;
- 9 percent decrease in precipitation;
- 10–50 percent reduction in surface water.

Activity 2: How Does Your Project Outline from Chapter 2 Relate to the Summarized Challenges Above?

Paste in your Field Assignment 2 project outline and your new climate stimuli side-by-side in a matrix in order to compare them. Go to Text Box 4.2 to see what this could look like.

Activity 3: How Do Your CBA Activities from Chapter 1 Relate to the Summarized Challenges Above?

Briefly describe in a short paragraph if your project challenges compare to the challenges described in the scientific summary in Activity 1 above. Go to Text Box 4.3 to see what this could look like.

Field Assignment 4 Step 2: Modifying the Project Outline Based on New Scientific Information

Activity 1: Do You Need to Modify Your Problems and Causes Based Upon What You've Learned?

If you feel that the CBA challenges from your original project outline are not in accord with the scientific information, this is where you can modify them. If this is the case, write a brief paragraph how you could modify your challenges. If your CBA challenges from your original outline are in accord with the new scientific information—well done, good job!

Box 4.2

Course Project Example

Table 4.1 Course project example: comparison of local knowledge with scientific information

Field Assignment 2 Project Outline	*Summary of climate-related stimuli and projected impacts*
Problems: • chronic diarrhea in children; • chronic under-nutrition. *New problem identified in Field Assignment 2 exercises* Insufficient income from agriculture *Causes*: Lack of knowledge of health, hygiene, and family nutrition Overall shortage of food and specifically for the four months preceding the corn harvest *New underlying cause related to climate change*: Unpredictable dates for the start and end of the rainy season, intermittent drought and erratic rainfall during the rainy season, flooding and extreme weather events have reduced crop harvests and access to water	*Near-term adverse climate change related stimuli:* Unpredictable dates for the beginning and end of the rainy season Decreasing precipitation Extended drought periods during the rainy season An increasing frequency of extreme precipitation events and an increase in frequency and intensity of extreme climatic events

Box 4.3

Course Project Example

The scientific climate stimuli are in accord with the CBA challenges identified by the community in Chapter 2 and are in support of their traditional climate knowledge.

These stimuli also represent/support underlying causes for other traditional development challenges in the outline: food shortages, insufficient agricultural income, and chronic malnutrition.

The idea is not to create a larger project that will attempt to solve all the new negative impacts that you discovered in your climate change research. The goal here is to find problems and underlying causes already in your project outline which are in agreement with the new scientific data or that need to be corrected based upon the new scientific data. You should keep it a small, simple project.

Write a brief paragraph about what you need to modify and how you propose doing it. Go to Text Box 4.4 to see what this could look like.

Box 4.4

Course Project Example

Activity 1: Do You Need to Modify Your Problems and Causes Based Upon What You've Learned?

The underlying cause related to climate change is:

- Unpredictable dates for the start and end of the rainy season, intermittent drought and erratic rainfall during the rainy season, flooding and extreme weather events have reduced crop harvests and access to water.

Chapter 3's scientific research showed that near-term climate stimuli and impacts concur with climate change causes in the project outline—so I do not need to modify that underlying cause.

The CBA cause in the project outline does not take into account a plan to address the long-term challenges. A long-term plan would need to be developed, but I believe that that is outside of the scope of this particular project. Developing a long-term plan would be a good idea for a second, follow-up project.

However, in anticipation of developing this long-term plan, I would like to add a new underlying cause to the project outline. This will create a bridge from this project to the follow-up project:

- A lack of knowledge of climate change and its near- and long-term impacts.

Revised Project Outline

Let's look at the local knowledge, and community-identified need combined with scientific knowledge.

Problems

- chronic diarrhea in children;
- chronic under-nutrition;
- insufficient income from agriculture.

Underlying Causes

- lack of knowledge of health, hygiene, and nutrition;
- overall shortage of food and specifically for the four months preceding the corn harvest.

Underlying Causes Related To Climate Change

- Unpredictable dates for the start and end of the rainy season, intermittent drought and erratic rainfall during the rainy season, flooding and extreme weather events have reduced crop harvests and access to water.

New Underlying Cause Based on Long-Term Climate Risks

- A lack of knowledge of climate change and its near- and long-term impacts.

Field Assignment 4 Step 3: Completing a Revised Project Outline Reflecting Local Need, Local Knowledge, and Scientific Climate Knowledge

Activity 1: Accurately Define Your Community's Local Climate Change Context

Almost as if you're writing a new problem statement for your problem outline, very briefly describe the exact climate change context that you have developed in this assignment. Hint: use the summaries that you wrote for step 1. Go to Text Box 4.5 to see what this could look like.

Box 4.5

Course Project Example

Activity 1: Accurately Define Your Local CC Context

100 subsistence farm families in four villages of Comalapa, Guatemala, are suffering from reduced crop harvests and access to water due to unpredictable dates for the start and end of the rainy season, intermittent drought and erratic rainfall during the rainy season, flooding, and extreme weather events—all exacerbated by a lack of knowledge of climate change and its near- and long-term impacts. These lead to a reduction in food security, increased malnutrition, and a decrease in agricultural income. These climate change challenges will intensify over the next 40 years; without adopting adaptation strategies, the community's suffering will increase.

Activity 2: Combining the Problem Statement from Field Assignment 2 with the New Context Statement

Copy the problem statement from the simple outline in Field Assignment 2, and the new climate change context statement from this chapter and paste them into a document as I've done below. Without creating redundancy or duplication, combine the two statements into one short statement. The simpler your problem statement is, the easier it will be to develop a theory of change. Go to Text Box 4.6 to see what this could look like.

Box 4.6

Course Project Example

Problem Statement from Field Assignment 2

Three hundred small children from 100 families in four Guatemalan villages are frequently ill with chronic diarrhea caused by little knowledge of health and hygiene, and are chronically under-nourished caused by little knowledge of nutrition and fewer than 12 months of food reserves. Their families suffer from insufficient agricultural income from reduced crop harvests and access to water due to unpredictable dates for the start and end of the rainy season, intermittent drought and erratic rainfall during the rainy season, flooding and extreme weather events. These challenges contribute to stunting and restrict the children's ability to attend and concentrate in school, leading to a reduction in their ability to develop and prosper as adults. These challenges also reduce the ability of adults to lead the productive, meaningful, prosperous lives they need to leave the cycle of poverty and contribute to the development of their communities.

Local Climate Change Context from Activity 1

One hundred subsistence farm families in four villages of Comalapa, Guatemala, are suffering from reduced crop harvests and access to water due to unpredictable dates for the start and end of the rainy season, intermittent drought and erratic rainfall during the rainy season, flooding, and extreme weather events—all exacerbated by a lack of knowledge of climate change and its near- and long-term impacts. These lead to a reduction in food security, increased malnutrition, and a decrease in agricultural income. These climate change challenges will intensify over the next 40 years; without adopting adaptation strategies the community's suffering will increase.

Activity 3: Inserting Revised Information into a Completed Project Outline

Paste your revised problem statement into your revised project outline. The finished problem statement should be concise and absolutely parallel to the outline. The statement should not introduce new ideas not found in the outline. Go to Text Box 4.7 to see how this looks or download the Example of Field Activity 4 to use as a template for this assignment.

Box 4.7

Course Project Example

Revised Problem Statement Incorporating the Community's Climate Change Context

Three hundred small children from 100 families in four Guatemalan villages are frequently ill with chronic diarrhea caused by little knowledge of health and hygiene and are chronically under-nourished caused by little knowledge of nutrition and fewer than 12 months of food reserves. Their families suffer from insufficient agricultural income from reduced crop harvests and access to water due to unpredictable dates for the start and end of the rainy season, intermittent drought and erratic rainfall during the rainy season, flooding, and extreme weather events—all exacerbated by a lack of knowledge of climate change and its near- and long-term impacts. These challenges contribute to stunting and restrict the ability of children to attend and concentrate in school. They also reduce the ability of adults to lead the productive, meaningful, prosperous lives they need to leave the cycle of poverty and contribute to the development of their communities.

Revised Project Outline

This revised project outline includes community-identified need combined with local and scientific climate change knowledge.

Problems

- chronic diarrhea in children;
- chronic under-nutrition;
- insufficient income from agriculture.

Underlying Causes

- lack of knowledge of health, hygiene, and nutrition;
- overall shortage of food and specifically for the four months preceding the corn harvest.

Underlying Causes Related to Climate Change

- Unpredictable dates for the start and end of the rainy season, intermittent drought and erratic rainfall during the rainy season, flooding, and extreme weather events have reduced crop harvests and access to water.
- A lack of knowledge of climate change and its near- and long-term impacts.

Chapter 4 Resources

Suggested Homework Assignment

The complete Field Assignment 4 homework to turn in will be:

1 a brief description of near-term and long-term climate change challenges that you discovered during your research;
2 a brief description of how your original project outline compares to the newly researched scientific information;
3 a brief description of how you would modify your program activities or introduce new activities to better address the challenges that you discovered in the scientific information;
4 your project outline with the proposed changes made;
5 a new context statement based upon the summary of the scientific information;
6 your problem statement from Field Assignment 2 combined with the new context statement;
7 your new problem statement, written up and inserted into your completed project outline.

Use the Example of Field Assignment 4 as the template for the assignment.

Course Downloads

Go directly to this book's webpage, TimMagee.net/field-guide-to-cba/ to download the following resources.

- Example of Field Assignment 4.

Recommended Resources

Website addresses change frequently. Simply enter this book's webpage for current links to resources, or enter the author's name, the organization's name and the document's name into your web browser to find the most current link.

CARE. *Community-Based Adaptation Toolkit*, CARE. Available at: http://www.careclimatechange.org/files/toolkit/CARE_CBA_Toolkit.pdf

Mitchell, T. and Tanner, T. *Adapting to Climate Change: Challenges and Opportunities for the Development Community*, Tearfund. Available at: http://www.tearfund.org/webdocs/website/Campaigning/policy%20and%20research/Adapting%20to%20climate%20change%20discussion%20paper.pdf

UNDP. *Designing Climate Change Adaptation Initiatives: A UNDP Toolkit for Practitioners*. UN. Available at: http://www.undp.org/content/undp/en/home/librarypage/environment-energy/low_emission_climateresilientdevelopment/designing-adaptation-initiatives-toolkit.html

Part II

Design

Part II focuses on designing a project and developing management tools both for presentation to donors and for implementing the project.

In Chapter 5, you will begin looking for solution-oriented programs and field activities that will address the challenges identified by your community and through your research on scientific climate data.

Next, you will research the activities to determine if there is an evidence basis that they have worked to solve the challenges represented in your project outline.

In Chapter 6 you will develop management and funding tools for the project. These will include a logical framework, a budget, a schedule, and a two-page compelling fact sheet.

5 Solution

Adaptation Activities

What Works in International Development?

In this chapter you will incorporate programs and activities into your simple project outline and create a project designed to reduce community vulnerability and strengthen resilience.

Chapters 1 and 2 have stressed the importance of community engagement and project ownership in developing a sustainable project. Another facet of sustainable projects is based upon using a series of building-block activities that have shown scientific evidence of having worked to provide effective solutions to challenges identified by your community. These are known as evidence-based best practices; we will encourage you to research and include these in your project.

Field Activities for Your Project Outline

What are some simple examples of adaptation activities that you could incorporate into your project? If your project is an agricultural project, and you're facing a water shortage challenge, successful ideas could be soil and water conservation activities such as building organic material into the soil. This allows the soil to act more like a sponge and hold water. Keeping mulch on the soil surface reduces evaporation. Channeling water to depressions in the soil can allow water to penetrate into the soil rather than run off. Low-tech drip irrigation systems can target water to the plants' roots.

In practice, community based adaptation (CBA) projects are very similar in appearance to traditional development projects. Traditional development activities are frequently used in adaptation-specific projects—but are focused on addressing a climate change challenge instead of a traditional development challenge.

In Field Assignment 5, after searching for potential activities, you will develop a simple theory about which activities will work best in your project, and then you will populate your project outline with these new activities. You will use traditional development activities for your project's traditional development challenges, CBA activities for your CBA challenges, and disaster risk reduction (DRR) activities for your DRR challenges. In order to develop a simple project that can be effectively managed, look for overlaps: can a single, individual activity address both a traditional development challenge and a CBA challenge at the same time? Two birds, one stone.

For example, in an earlier example of a village spring that dried up, one proposed solution had been to restore the watershed that originally had charged the spring through a reforestation project. Is this good development, good adaptation, good mitigation, or good disaster risk reduction? Or is this a good solution that was developed that addressed four sectoral problems at the same time?

What is also important in this assignment is to continue to reinforce the fact that your project is rooted in the community. So the last step in this chapter will be to return to your community to share your expanded project outline in an effort to keep them engaged as partners in the process and to gain their feedback.

Your community has its own prioritization of its needs. If you address the needs that they originally voiced in the design of your project, you will be continuing the process of community buy-in. Climate change is only one of a range of problems that communities face and it is unlikely that activities focusing only on climate-related risks will reflect the community's full priorities. If you do your job well, the community should feel that you're working on their project for them. If they agree with the project concept, this will help confirm their sense of ownership: this is their project and you are simply a short-term expert who has come to offer assistance in developing it.

Chapter 5 will help you accomplish five steps in the development of your project. You will do the following:

Step 1: undertake research into solution-based project field activities.
Step 2: compile a program and activity solution list.
Step 3: verify field activity effectiveness through scientific evidence.
Step 4: present the project design to the community for feedback.
Step 5: assess your NGO's project expertise.

What you'll need:

- access to an Internet browser or university search engine;
- meeting with representatives of the previous workshops.

Timeframe:

- 4 hours:
    - 1 hour online research;
    - 1 hour to incorporate programs into outline;
    - 2 hours for the community meeting.

Field Assignment 5 Step 1: Researching Solution-based Project Activities

In this step you will be developing a theory of how you plan to address the challenges in your project. You concluded Field Assignment 4 with a set of community-identified and scientifically identified challenges organized into a simple project outline. Now is your opportunity to develop a theory of how to solve these challenges, and to begin exploring specific activities that will prove your theory.

In the development world it is called a theory of change; it's your theory of what changes in behavior, changes in infrastructure, or changes in capacity will need to be realized to solve the problem. Your theory of change will include the activities needed to address the underlying causes of the problems listed in the project outline. You will also draft a goal statement that reflects the positive results of your theory of change.

Activity 1: Searching for Project Activities

On a sheet of paper, make a list of your community-identified problems and underlying causes from your simple project outline with enough space beneath each one to write down ideas for potential solutions in the form of programs and field activities.

Approach colleagues in your office, use your own experience, and explore the Internet for ideas. Be sure and use this book's webpage as a resource. Visit Chapter 5 Resources. Make a note of the best two Internet resources you discover with program and activity ideas that are appropriate for your project. Write a short paragraph that summarizes programs or activities that will support your theory of change. Go to Text Box 5.1 to see what this could look like.

Box 5.1

Course Project Example

Activity 1: Searching for Project Activities: A List of Websites that You Consulted for Potential Activities

I searched the Internet for development resources on food security, nutrition and health, and hygiene sites, and I checked the Chapter 5 Resources. I found the following sources for the health, hygiene, and nutrition components of my project:

> Burgess, A. and Glasauer, P. (2004) *Family Nutrition Guide*, Rome: Food and Agriculture Organization of the United Nations. Available at: http://www.fao.org/docrep/007/y5740e/y5740e00.HTM (accessed 21 April 2012).
>
> Food and Agriculture Organization of the United Nations (2001) *Improving Nutrition through Home Gardening: A Training Package for Preparing Field Workers in Africa*, Rome: FAO. Available at: http://www.fao.org/docrep/003/x3996e/x3996e00.htm (accessed 21 April 2012).
>
> UNDP-WHO (1994) *Food, Water and Family Health: A Manual for Community Educators*, UNDP. Available at: http://www.who.int/water_sanitation_health/hygiene/settings/wsh9204.pdf (accessed 21 April 2012).

These papers discuss the success of home gardens in increasing food security and food diversity, but also discuss the importance of nutritional education as a part of developing these gardens. The UNDP paper discusses the importance of properly disinfecting water and also the importance of clean, safe water storage. I chose what I thought were the best activities to address the problems and their underlying causes, and then wrote them down beneath the problems within the problem list.

Adaptation Component

I searched the Internet for development resources on adaptation and on the impacts of erratic rain and drought on agricultural productivity. I also checked Chapter 5 Resources. I found the following online resources on agriculture and adaptation:

Kato, E., Ringler, C., Yesuf, M. and Bryan, E. (2009) *Are Soil and Water Conservation Technologies a Buffer Against Production Risk in the Face of Climate Change? Insights from Ethiopia*, IFPRI Discussion Paper No. 871. Available at: http://www.ifpri.org/sites/default/files/publications/rb15_17.pdf (accessed 21 April 2012).

The position of this paper is that soil and water conservation technologies have significant impacts in reducing production risk as part of a climate proofing strategy. The results also show that one-size-fits-all recommendations are inappropriate, given the differences in agro-ecologies and other factors. Therefore, the performance of these technologies is location- and context-specific. This paper also details 10 activities that can be incorporated as adaptation activities into agricultural systems at risk.

Food and Agriculture Organization of the United Nations (2010) *"Climate-Smart" Agriculture*. Online. Available at: http://www.fao.org/fileadmin/user_upload/newsroom/docs/the-hague-conference-fao-paper.pdf (accessed 21 April 2012).

This paper discusses the formidable task of increasing agricultural production in the face of both a growing world population and negative impacts from climate change. It argues that the use of improved and conservation agricultural practices can help farmers adapt to a changing climate and at the same time provide carbon sequestration in the soil of farmers' fields. These practices will increase productivity, resilience, and will enhance food security. The practices they suggest include soil and nutrient management, water harvesting, pest and disease control, resilient ecosystems, genetic resources, improved post-harvest processes, conservation agriculture, and agroforestry.

From reading through these sources and from ideas uncovered in my meetings, I chose what I thought were the best adaptation programs and activities to address the adaptation problem, and then put them in the problem list.

Field Assignment 5 Step 2: The Program and Activity Solution List

Activity 1: Expanding Your Project Outline

Download the Example of Field Assignment 5 and use that as your assignment template. Simply copy and paste your new field activity ideas from your project outline right over the top of what is written in this formatted document.

Follow the order of your project outline and begin with the first problem and underlying cause. Make a decision about a title for a program that would address the first underlying cause. In order to keep your project simple, try and limit yourself to only one or two programs as a solution to one underlying cause. Write in the name of the underlying cause adjacent to the program title as in the following example.

Now add the activities from your new list of activities that you discovered beneath the program title in an outline format. The activities should be related to the program and when added together, the individual activities should add up to equal the full intent of the program. If sequential, the activities should be arranged in chronological order.

So you should have an outline with this format:

Project Outline: Problem List Combined with Potential Programs and Solution-Oriented Activities

[Problem 1] Chronic diarrhea in children
Health and Hygiene Program [Solution to underlying cause: Lack of knowledge of health and hygiene]:
[Activity 1] Hand washing workshop and follow-up.
[Activity 2] Point of use water purification workshop and follow-up.

You should also think through the sustainability of the program. Are there activities in the program—such as setting up community water use committee—that will ensure program continuation after you've handed over the project to the community? Are there capacity building workshops for community members so that they can continue the implementation of program activities?

Notice in the example that I copied and pasted in the problem statement from Chapter 4. I also annotated the problem statement, the project outline, and the goal statement so that you can see how they interrelate and that they are all parallel to each other. Please do the same with yours. It will become obvious how important this is in the coming chapters.

Activity 2: Goal Statement

If your problem statement is clear, concise, and simple, you will have a much easier time developing your goal statement. Your project's goal statement will be phrased to be a positive reflection of your problem statement (somewhat like mirror images of each other) and at the same time offer a general strategy for a solution (your theory of change).

To create your goal statement, simply copy your problem statement and paste it below your project outline—and without changing it structurally—make the negative statements into positive statements by changing the negative words. It needs to be left clean and simple and be an exact, parallel reflection of the problem statement. Don't be tempted to add new information or expand upon project ideas. For example, if one of your problems says that "Community members don't have enough food to eat for the four months before the harvest," then your goal statement might say:

> Community members will be able to enjoy food security 12 months of the year.

Then after that positive statement, include one of your proposed programs such as:

> Through a family garden and nutrition program.

This addition of the proposed programs to your goal statement details your theory of change for addressing the community's challenges. Go to Text Box 5.2 to see what this could look like, or download the Example of Field Assignment 5.

Box 5.2

Course Project Example

Field Assignment 5 Step 2: The Program and Activity Solution List

PROBLEM STATEMENT

[Problems and underlying causes] (1) 300 small children from 100 families in four Guatemalan villages are frequently ill with chronic diarrhea caused by little knowledge of health and hygiene and (2) are chronically under-nourished caused by little knowledge of nutrition and less than 12 months of food reserves. These families also suffer from (3) insufficient agricultural incomes from (4) reduced crop harvests and access to water due to unpredictable dates for the start and end of the rainy season, intermittent drought and erratic rainfall during the rainy season, flooding, and extreme weather events—all exacerbated by (5) a lack of knowledge of climate change and its near- and long-term impacts. They contribute to [Negative impacts] (a) stunting and restrict the ability of children to (b) attend and concentrate in school. These challenges also (c) reduce the ability of adults to lead the productive, meaningful, prosperous lives they need to leave the cycle of poverty and contribute to the development of their communities.

Project Outline: Problem List Combined with Potential Interventions/ Activities/Solutions that I Chose

[Problem 1] Chronic diarrhea in small children

Health and Hygiene Program [Solution to underlying cause: Lack of knowledge of health and hygiene]

[Activity 1] Hand washing workshop and follow-up

[Activity 2] Point-of-use water purification system workshop and follow-up

[Problem 2] Chronic under-nutrition

Family garden and nutrition program [Solution to underlying causes: Lack of knowledge of family nutrition; Overall shortage of food and specifically for the four months preceding the corn harvest]

[Activity 1] Workshop and follow-up in family nutrition and home garden planning for nutrition

[Activity 2] Forming beds and planting seeds workshop and follow-up

[Problem 3] Insufficient income from agriculture

Agricultural Income Generation Program [Solution to underlying causes: Unpredictable dates for the start and end of rainy season, intermittent drought and erratic rainfall during the rainy season, flooding, and extreme weather events have reduced crop harvests and access to water]

[Activity 1] Facilitate the organization of a community based farmers association

[Activity 2] Survey local/regional businesses or markets that buy and sell agricultural produce in order to determine products they need on a routine basis

[Activity 3] Establish a market link and ask the businesses for their support in training programs and inputs for farmers to grow the products they need

[Activity 4] Using this input, launch training workshops on improved agricultural practices for these new crops and markets

Program Related to Climate Change but also in Support of Traditional Development Challenges Above

Climate-Smart Agricultural Practices Program [Solution to underlying causes: Unpredictable dates for the start and end of rainy season, intermittent drought and erratic rainfall during the rainy season, flooding and extreme weather events have reduced crop harvests and access to water; Overall shortage of food and specifically for the four months preceding the corn harvest]

[Activity 1] Identify expert specialist/extension agent in soil, water, and agriculture to design and facilitate participatory workshops

[Activity 2] Participatory mapping and identification of local soil, water, and crop challenges

[Activity 3] Consciousness-raising workshop on soil and water conservation and improved agricultural practices

[Activity 4] Farmers workshop on soil restoration and conservation techniques—including composting and mulching

[Activity 5] Farmers workshop on water conservation and management techniques—including water harvesting techniques

[Activity 6] Farmers workshop and follow-up on early maturing and/or drought resistant crops for adapting to climate variability

Climate Change Awareness Program [Solution to underlying causes: A lack of knowledge of climate change and its near- and long-term impacts]

[Activity 1] Community based workshop and survey to identify their knowledge of climate change and its potential near- and long-term challenges

[Activity 2] Consciousness-raising workshops about climate change, its near- and long-term impacts, and the need for a long-term adaptation plan

[Activity 3] Discussion and community prioritization of developing a long-term adaptation plan as a follow-up project to this project

Goal Statement

This is an exact positive reflection of the Problem Statement; paste in your problem statement and simply make it positive.

Three hundred small children from 100 families in four Guatemalan villages will be able to [Underlying causes to problems as if they have been solved] (1) enjoy better health through a health and hygiene program, and (2) have improved nutrition and 12 months of food security per year through a family garden and nutrition program. These families will also (3) enjoy increased agricultural incomes through an agricultural income generation program, and (4) increased crop harvests and access to water through a climate smart agricultural practices program and through a (5) climate change awareness program. These opportunities will contribute to [Positive impacts] (a) improved growth and development for children, (b) better school attendance and performance—and will also increase (c) the ability of adults to lead the productive, meaningful, prosperous lives they need to leave the cycle of poverty and contribute to the development of their communities.

What you have created here is the basis for a full project. This is the beginning of a logical framework—a log frame. You're going to use this version of the completed the project outline just as it is for the rest of the chapters in this book.

Mainstreaming: Integrating Adaptation Activities and Traditional Development Activities

Stepping back a pace from the outline, notice that there are three fairly traditional development problems in the outline: children with poor health, chronic under-nutrition, and insufficient agricultural income. There are also two climate change challenges: climate variability and a lack of knowledge about climate change.

Look at how the adaptation programs interrelate with the traditional programs. The climate smart agricultural program not only addresses the adaptation challenges, it also works in support of addressing chronic under-nutrition and insufficient agricultural income generation. The climate smart agricultural program bridges traditional development and community based adaptation. Also, the three traditional development programs can be viewed as supporting community based adaptation through reducing vulnerabilities, strengthening the community, and building resilience to climate change.

Field Assignment 5 Step 3: Verifying Activity Effectiveness through Evidence

In this step you will be investigating if there is a scientific basis that your proposed theory and activities have worked on other projects.

Suppose that you are a mother whose children are suffering, and an unknown organization came to you with a plan to help your children. Wouldn't you want that plan to work? Suppose that you are a donor hoping that your donations will fulfill some need. Wouldn't you want your donations to have an impact? Suppose that you were a local NGO hoping to improve the lives of your people. Wouldn't you want to be successful?

Today it is acknowledged that development programs have not kept up with increasing need. One of the very simple reasons is that organizations copy what other organizations are doing without investigating if the programs are working and if they are having any lasting impact. There is an extraordinarily simple solution to this: do a bit of research to see if any scientific studies have been done on the effectiveness of a proposed activity. The more information that you can find about project activities that have shown evidence of having worked to solve your problem, the greater the likelihood that:

- your theory of change will be a good one;
- your project will have long-term positive impact.

Be careful, because your research might lead you to studies on activities which you are enthusiastic about, but it may not be an activity that shows evidence of having worked. So, at this early stage, before you fall in love with your project programs and activities, you have the opportunity to research whether there is a basis in scientific evidence that they work.

What is a Scientific, Peer-Reviewed Paper?

After conducting a study and writing a scientific paper, a scientist may submit the paper to a journal. The journal will ask scholars in the same scientific field—the author's peers—to review the paper in an effort to determine the quality of the scientific research, scholarship, and suitability for publication. This is known as a peer-reviewed study. It is these papers that you are looking for. If you are connected to a university, these studies will be easy to find. Universities subscribe to scientific journals and to search engines which are programmed to find scientific studies.

These studies will provide you with an abstract that will tell you in one paragraph the results of the study. The main body of the study will give you more detailed information on why the activity did or did not work and under what circumstances. The circumstances are very important because you may discover that your proposed activity may not work in some situations—but will work perfectly in a different situation. Make sure that your proposed activity has shown evidence of having worked in your situation. What if you find evidence that your proposed activity doesn't work? Determine if the problem only entails a minor modification to the activity (which you may be able to do). Otherwise, search again for new activities that will work—and modify your project outline accordingly.

Activity 1: Researching Scientific Documents for Evidence of the Effectiveness of Your Project Activities

In this step, you will be researching two activities. Pick activities you plan to use in your project that are simple ones and that can be easily described. In your search engine you can ask questions just as if you are asking them of another person. For example, in my project, I am looking for information on improving children's health. So in my search engine, I can enter "What works in reducing diarrhea in children in developing nations?"

Using keywords like "international development" and "developing nations," will get you out of mainstream news and into real development information. You can also type in words like "abstracts" and "executive summaries" to point you towards scientific documents. When you find a document, it may be a "synthetic study" or a "literature review," which has analyzed a large number of primary research documents. These will give you the most useful results for your project. You are not looking for informational documents found on websites that have not been through an evaluation by a team of scientists. These are called 'gray literature' and are not acceptable as scientific evidence.

Once you've found one or two of these scientific documents, act like a scientist: without putting your own desires and feelings into the interpretation of the document, write a short, one paragraph summary that will indicate whether or not the activity that you have selected for your project appears to work, or not. Do this for two of your project activities. If a document indicates that the activity you chose to solve your project challenge does not work—that's good news too; it means that you won't waste time and money on an activity which is not going to perform. Go to Text Box 5.3 to see what this could look like.

Box 5.3

Course Project Example

Activity 1: Researching Scientific Documents for Evidence of the Effectiveness of Your Project Activities

I focused on finding scientific papers on the following two of my activities to see if they had shown evidence of solving my project challenge. I searched the Internet, and used Chapter 5 Resources.

Project Activity 1. Family gardens and nutrition. Key words: Increasing food security and nutrition with family gardens and small animal production in development.

Bhattacharjee, L., Phithayaphone, S., and Nandi, B. K. (2006) *Home Gardens Key to Improved Nutritional Well-Being*, Bangkok: FAO. Online. Available at: ftp://ftp.fao.org/docrep/fao/meeting/011/ag101e/ag101e00.pdf (accessed 21 April 2012).

English R. M., Badcock, J. C., Giay, T., Ngu, T., Waters, A-M., and Bennett, S. A. (1997) Effect of nutrition improvement project on morbidity from infectious diseases in preschool children in Vietnam: comparison with control commune, *BMJ* 315, 1 November.

Iannotti, L., Cunningham, K., and Ruel, M. (2009) *Improving Diet Quality and Micronutrient Nutrition: Homestead Food Production in Bangladesh*, IFPRI. Online. Available HTTP: http://www.ifpri.org/sites/default/files/publications/ifpridp00928.pdf (accessed 21 April 2012).

Summary Paragraph

These studies show that home gardens can provide 60 percent of leafy vegetables, and between 20 percent and 50 percent of all fruits and vegetables consumed by households. Home gardening families as a rule spend less on food than non-gardening families. Improved nutrition boosts the body's immune system protecting children against disease and can reduce the number of children with diarrheal infections from a total of 18 percent of the children down to 5 percent of the children. One study showed that after six months of a vegetable garden project, the number of malnourished children decreased from 23 percent in the communities to 16 percent and the number of severely malnourished children decreased from 9.5 percent to 2 percent. The studies all emphasized that vegetable gardens needed to be combined with nutrition education so that mothers could make sure that they were growing a variety of vegetables and fruit rich in vitamins and minerals, especially vitamin A. The gardens were a good source of protein through eggs and small animal production. The studies also concluded that even a small, 25 square meter garden can have a positive impact on nutrition, health, and increased incomes.

Project Activity 2. Hand washing for improving children's health. Key words: What works in reducing diarrhea in developing nations?

Waddington, H., Snilstveit, B., White, H., and Fewtrell, L. (2009) *Water, Sanitation and Hygiene Interventions to Combat Childhood Diarrhoea in Developing Countries*, 3ie. Online. Available HTTP: http://www.3ieimpact.org/admin/pdfs2/17.pdf (accessed 21 April 2012).

Zwane, A. P. and Kremer, M. (2007) What works in fighting diarrheal diseases in developing countries? A critical review. *World Bank Research Observer* 22(1): 1–24.

Summary Paragraph

Randomized control studies have shown that hand washing can reduce diarrhea in children in developing nations by between 30 percent and 53 percent. Hands can pick up pathogens that cause diarrhea in the latrine, by washing hands in infected water, by touching another person's hands, and by touching the ground where someone has tracked fecal matter. Hand washing with soap is the number one prevention against the spread of person-to-person infection. Hand washing reduces the spread of germs that cause diarrhea, respiratory illness, and skin infection.

Conclusion

I was lucky that scientific evidence supported the fact that two of my project activities have shown evidence of having worked in a situation similar to my community's situation.

Field Assignment 5 Step 4: Community Ownership: Feedback, Input, and Engagement

Will the Community Buy into It?

By now, your project is beginning to take shape. You last saw your community after completing the capacity and vulnerability assessment. Although they had an idea of the kind of project you might be developing, your project has progressed tremendously since then. It would be a very good idea to return to the community, just for a short meeting, to let them see how the project has evolved before you invest more time in it.

Meeting with the community at this stage will let you see if you've accidently designed any cultural taboos into the project. But it also has a larger purpose: you are seeking their comments. This will cause them to feel a sense of increasing ownership of the project. They will perceive you not as an outsider presenting a canned project to them, but as someone who's working on their behalf and following their suggestions.

This buy-in is absolutely paramount to the long-term sustainability of the outcomes of the project. If they like what they see as the project design evolves, and truly feel that it was based upon their ideas, they are going to protect and take care of the outcomes long after you're gone.

Activity 1: Feedback from the Community

In getting feedback from the community, all that you will need is to organize a meeting lasting 30–60 minutes. You can meet with the whole group that attended the previous workshops, or you can just meet with a few representatives—whichever is easiest for both you and for them. Discuss your Assignment 5 Project Outline and Project Goal with them. At this stage don't bring out specialized details. Listen to their reactions, answer any questions that they may have and affirm their feedback.

Box 5.4

Course Project Example

Activity 1: Feedback from the Community

I met with six participants of the last workshop and discussed the project with them. This particular group has been my most consistent set of contacts over the past few months as the project has been developing. On the one hand, they seem to have the best grasp of the project, on the other, they seem to have the community's interest in mind. So they're a good group and I enjoy working with them.

They seemed happy that I was pursuing the needs that they had identified and had come up with practical activities. They understood all of the activities. As I suspected, there weren't any surprises since they have seen this project slowly unfold over several months. They seem happy that I have stayed with this project for so long, and have been consistent in pushing the project forward and keeping them informed. In these meetings I try to represent myself as someone who's just temporarily in support of what they're trying to do—this is, after all, their project.

Field Assignment 5 Step 5: Assessing NGO Expertise

Step 5 of this assignment is to make a list of project activities where you or your NGO have insufficient experience or expertise. This book is about "What works in development?"—so we want to make sure that each activity has someone in charge who has the expertise to make it successful—and sustainable.

- What expertise do you need to implement your project?
- What expertise do you have?
- What expertise do you need to acquire?
- Where can you get it?

In the next field assignment, you will take your project to a donor for their pre-submittal feedback. Donors can be unusually helpful in fine-tuning a project; they have seen a lot of projects and have a sense of what works and what doesn't.

Two of the things that they can be quite picky about are experience and expertise. I was in a meeting with an NGO who approached a donor with a point-of-use water filter project; the NGO was very excited about their project design. The donor asked about the NGO's experience in working with water filters in communities—and was surprised to find out that their expertise was actually in reforestation projects—not in water filters. Needless to say, they didn't get the grant award.

Activity 1: List of Project Consultants and Working Partners

In order to help you avoid an embarrassing situation like this, Step 5 of this assignment is to make a list of aspects of your project activities where you and your NGO have insufficient experience. Make a three-column table:

1 The first column is the list of activities where you have insufficient experience.

2 The second is your solution to finding expertise for each activity. List one of three things; (a) Hire a short-term expert consultant; (b) partner with an experienced NGO; or (c) hire a specialist for the duration of the project.
3 The third column is a list of who these consultants or partners could be or where you could find them.

This short process will ensure that you are not caught unprepared in a donor meeting, and will come in useful in Field Assignment 6 when you are developing your project budget.

For an NGO to submit a project to a donor that includes an activity in an area where they don't have expertise is perfectly normal. Frequently NGOs with complementary expertise partner with each other, frequently NGOs hire expert consultants, and frequently NGOs hire specialists for the duration of the project. This just needs to be clearly spelled out in your proposal and budget for the donor to be satisfied. Go to Text Box 5.5 to see what this could look like.

Box 5.5

Course Project Example

Activity 1: List of Project Consultants and Working Partners

Table 5.1 Course project example: assessing NGO expertise

Activity requiring expertise	*Consultant or NGO partner?*	*Who/Where?*
Water purification	NGO water filter partner	Local NGO that makes, supplies, and trains communities in the use of water filters
Family nutrition	Nutrition consultant	In-country consultant who understands local cultures, and teaching nutrition I will contact a local university or the Ministry of Health
Home gardens	Agricultural consultant or the Ministry of Agriculture	The local Ministry of Agriculture will be able to introduce me to a consultant who knows the cultures, the different in-country ecosystems, and appropriate plants

I had a heart-to-heart look at my project outline and realized that I was neither an expert in water filters nor nutrition—and even though I am an expert in home gardens, I am not 100 percent familiar with what would be appropriate to grow in Guatemala. So I listed those three things in column one.

I realized that I have a colleague whose NGO manufactures reputable water filters and provides training for the communities as part of their service. Partnering with them would bring the expertise I need.

I have worked on projects with a number of nutritionists in Guatemala and they really know their stuff. They will know exactly what to do within the different cultures in Guatemala as well as what nutritious things are acceptable to indigenous communities. The nutritionists not only can help us choose nutritious fruits and vegetables, they can work with mothers on gaining a better understanding of family nutrition, meal planning, and cooking.

I have also worked with agronomists who have the knowledge that bridges the gap between pure nutrition and planting a family garden. They will be able to help train our field staff to lead workshops, work with us on making the final selection of crops, and keep an eye on the progress of the workshops and the growth and maintenance of the gardens.

Chapter 5 Resources

Suggested Homework Assignment

The complete Field Assignment 5 homework to turn in will be:

1. a list of colleagues and websites that you consulted for potential activities;
2. an expanded project outline that includes the potential programs and activities/solutions that you have chosen your goal statement;
3. a list of one or two scientific papers about two of your proposed project activities;
4. a short paragraph summarizing the findings from the scientific papers for both activities;
5. a short paragraph summarizing a meeting with your community to discuss the progress of the project;
6. a three-column table with a list of activities requiring expertise and the identification of potential expert partners.

Course Downloads

Go directly to this book's webpage, TimMagee.net/field-guide-to-cba/ to download the following resources.

- Example of Field Assignment 5;
- CBA Specific Recommended Links;
- Solution-Oriented Program Templates for Challenges and Relief, Development and Adaptation.

6 Project

Management and Funding Tools

Developing a Set of Project Management Tools

In this chapter you will develop the management and funding tools that you will need to launch the project. Your project outline will be transformed into a logical framework (log frame) complete with problem, goal, and impact statements, a detailed budget, a schedule, and a concise project description—all for presentation to a donor. The development of each of these documents will be taken in simple steps using downloadable templates and examples.

How Will You Organize Your Ideas for Presentation to a Donor?

Using information from your completed assignments, you will write a two-page compelling project fact sheet. You will then polish, print, and bind your field assignments, meet with a colleague for input, and then make an appointment with a donor to present your project for feedback.

User-Friendly Management Documents

The documents presented in this chapter are very simple examples of log frames, budgets, schedules, and fact sheets.

1. If you haven't developed tools like this before, their simple format will ease your learning process.
2. If you have developed documents like these before, you will find that these simple examples are very easily adaptable to the wide variety of document formats (such as log frame or budget configurations) that different donors and NGOs use.
3. If your organization has a specific (and perhaps more sophisticated) format for these documents, please feel free to use those in place of the examples provided.

This chapter will help you accomplish five steps in the development of your project. You will:

1. develop a logical framework;
2. develop a detailed budget;
3. develop a schedule;
4. write a compelling fact sheet about your project;
5. request feedback from a donor.

What you'll need:

- access to a computer and e-mail;
- a meeting with a colleague;
- a meeting with a donor.

Timeframe:

- 10 hours:
  - 5 hours for the log frame, budget, and schedule;
  - 3 hours for the fact sheet;
  - 1 hour for the meeting with a colleague;
  - 1 hour for the meeting with a donor.

Field Assignment 6 Step 1: Developing a Logical Framework

Log Frames

The logical framework (log frame) is a powerful tool that aids in project planning, budgeting, management, and monitoring and evaluation. A log frame is a planning tool that helps to organize your project so that you can achieve your project goals—and a management tool that helps you to manage the project so that it will run efficiently and achieve the desired impact. In working with donors and stakeholders, your log frame will allow you to communicate to others exactly what you are trying to accomplish, and convey to them if you are making progress toward your goals.

Activity 1: Inserting Your Project Outline into a Simple Matrix

In this first activity you're going to take your project outline from Field Assignment 5 that contains your problem statement, project outline and goal statement, and insert these components into a simplified matrix: this is the first step in building a logical framework. This will only take 30 minutes to do but you will be able to see an immediate change in the organization of your project:

1. Open your final project outline from Field Assignment 5.
2. Download a simple matrix called Example of Field Assignment 6 Log Frame 1 to use as a template for your own project. The matrix has already been filled in with the course's example project. You will copy and paste the different components of your project outline into the matrix as illustrated by our example matrix. Please do not change your project from Assignment 5—the outline and the log frame will need to remain parallel. Leave the empty cells, columns, and rows found in the matrix empty; you will use them in the next activity.
3. Choose a working title for your project and enter it into the top of the matrix.
4. Paste your problem statement and the goal statement into the appropriate cells. You can see that the matrix is annotated, allowing you to see that the problem statement and the goal statement are parallel to the sub-goals below.

5 Copy the names of your different programs from your project outline and paste them into the sub-goal program title cells: Sub-Goal 1 has the short program title from the project outline: Health and Hygiene Program.
6 Paste the sentence from your positive goal statement that relates to the sub-goal program title into the cell next to the sub-goal program title: this will be the description of the sub-goal.

The sub-goals should be sub-components of your main goal and, when added together, should fulfill the main goal. For example, Sub-Goals 1 and 2 equal the solutions 1 and 2, noted in the main goal. Make sure that the problem statement, goal and sub-goals remain absolutely parallel to each other.

For example, one of the course example's problems is "300 small children from 100 families in four Guatemalan villages are frequently ill with chronic diarrhea caused by little knowledge of health and hygiene." This is positively reflected in the goal statement: "300 small children will experience a reduction in chronic diarrhea through a health and hygiene program." Paste your equivalent statement into your Sub-Goal 1 description cell.

1 Copy and paste in your program activities below their appropriate sub-goals next to the cells labeled Outputs. In this exercise you are converting the activities from the simple outline into outputs in the log frame. Your first activity (Activity 1) from your first program would therefore go into the output description cell next to Output 1.1, and so on. You may need to add more rows to the table if you have additional sub-goals and outputs.
2 Download the Example of Field Assignment 6 Log Frame 1 as a template for this assignment, or go to Text Box 6.1. The examples of the field assignments in the book are condensed versions of the complete project. The download examples of the field assignments contain the fully detailed project.

Box 6.1

Table 6.1 Course project example: Field Assignment 6 Log Frame 1

100 Maya families will enjoy improved health, food security, and incomes from a health, family garden, livelihoods, and climate-smart agricultural project	
Problem Statement	[Problems and underlying causes] (1) 300 small children from 100 families in four Guatemalan villages are frequently ill with chronic diarrhea caused by little knowledge of health and hygiene and (2) are chronically under-nourished caused by little knowledge of nutrition and fewer than 12 months of food reserves. These families also suffer from (3) insufficient agricultural incomes from (4) reduced crop harvests and access to water due to unpredictable dates for the start and end of the rainy season, intermittent drought and erratic rainfall during the rainy season, flooding, and extreme weather events—all exacerbated by (5) a lack of knowledge of climate change and its near- and long-term impacts. They contribute to [Negative Impacts] (a) stunting and restrict the ability of children to (b) attend and concentrate in school. These challenges also (c) reduce the ability of adults to lead the productive, meaningful, prosperous lives they need to leave the cycle of poverty and contribute to the development of their communities.

Goal Statement	[Goal Statement]: 300 small children from 100 families in four Guatemalan villages will be able to [Underlying causes to problems as if they have been solved] (1) enjoy better health through a health and hygiene program, and (2) improved nutrition and 12 months of food security per year through a family garden and nutrition program. These families will also (3) enjoy increased agricultural incomes through an agricultural income generation program, and (4) increased crop harvests and access to water through a climate-smart agricultural practices program and through (5) a climate change awareness program. These opportunities will contribute to [Positive Impacts] (a) improved growth and development for children, (b) better school attendance and performance—and will also increase (c) the ability of adults to lead the productive, meaningful, prosperous lives they need to leave the cycle of poverty and contribute to the development of their communities.
(1) Sub-Goal 1 (Objective) [sub-components of the main goal, which when added together describe the main goal]	
Health and Hygiene Program	300 children will experience a reduction in chronic diarrhea through a health and hygiene program
Output 1.1	Hand washing workshop and follow-up
Output 1.2	Point-of-use water purification workshop and follow-up
(2) Sub-Goal 2 (Objective) [sub-components of the main goal, which when added together describe the main goal]	
Family Garden and Nutrition Program	100 families and their 300 children in four villages will be able to enjoy better overall nutrition and 12 months of food security per year through a family garden and nutrition program
Output 2.1	Workshop and follow-up in family nutrition and home garden planning for nutrition
Output 2.2	Forming beds and planting seeds workshop and follow-up
(3) Sub-Goal 3 (Objective) [sub-components of the main goal, which when added together describe the main goal]	
Agricultural Income Generation Program	100 families and their 300 children in four villages will be able to enjoy sufficient agricultural incomes through an agricultural income-generation program
Output 3.1	Organize a community based Farmers Association
Output 3.2	Survey local/regional businesses or markets that buy and sell agricultural produce
(4) Sub-Goal 4 (Objective) [sub-components of the main goal, which when added together describe the main goal]	
Climate-Smart Agricultural Practices Program	100 families and 300 children in four villages will be able to enjoy better overall nutrition, increased crop harvests and access to water, and have a greater resilience to a changing climate through a climate-smart agricultural practices program
Output 4.1	Design participatory workshops
Output 4.2	Participatory workshops in mapping of soil, water, and crop challenges

(5) Sub-Goal 5 (Objective) [sub-components of the main goal, which when added together describe the main goal]

Climate Change Awareness Program	100 families and 300 children in four villages will be able to have a greater resilience to a changing climate through a climate change awareness program
Output 5.1	Workshop and survey to identify their knowledge of climate change
Output 5.2	Consciousness-raising workshops about climate change, its near- and long-term impacts, and the need for a long-term adaptation plan

Completion of the Log Frame

The last activity was to place your goals, sub-goals, and outputs into a planning matrix. In this activity you're going to begin the process of turning that matrix into a full-fledged log frame.

Activity 2: Measurable Sub-goals

Download the Example of Field Assignment 6 Log Frame 2. You'll see that it's the same matrix that you have just completed, but that a number of the previously empty cells have new information.

In order to assess whether the programs and activities that you chose for the project are working, you need to have something measurable. Start by making your sub-goals measurable. From the research that you did on your activities in Chapter 5, you may have found levels of success written up in the abstracts of the scientific papers that you read. For example, the scientific studies that I found indicated that hand washing could reduce diarrhea in children by between 30 and 53 percent and that ceramic water filters could reduce diarrhea by between 40 and 50 percent.

At this early stage of project design you do not want to overstate the successes that you hope to achieve. So I was conservative and said that 50 percent of the children in my project will be free from chronic diarrhea by the end of the project.

In a similar fashion, results on family gardens show that they can improve nutrition by 50 percent—and so I'm using that number for my expected improvement as well. You can look at the example log frame template and see that I have added these percentages to my sub-goal descriptions. Please do the same to your project log frame.

Activity 3: Expansion of Output Descriptions

In as much as the sub-goals need to add up to fully describe the project goal, the outputs also need to add up to fully describe the sub-goals. So you need to expand your definitions of the outputs to make sure that they will fully describe the sub-goals. You also need to make the descriptions of your outputs very specific and measurable.

The first output for the example log frame said "hand washing workshop and follow-up" has now changed to say "100 families in four communities participated in workshops on the benefits and practice of hand washing and received six months of follow-up." This describes the output as if it has happened. It is something measurable. And it relates to concrete elements within the sub-goals: 100 families from four communities. It also gives a timeframe: six months.

Activity 4: Detailed Activities

Just as outputs add up to fulfill the sub-goals, you must now define specific activities which will add up to fulfill the outputs. The first output is a workshop on hand washing. Think creatively. You may, for example, need to develop a lesson plan and collect materials. You will need to arrange meetings for the workshops with the communities. You will need to present the workshops. And then you will need to provide follow-up for six months in order to encourage the community to adopt hand washing in their daily lives. You can see these four activities in the example beneath the output. They have also been numbered. Please do the same to your project log frame, and then complete the log frame by filling in your other activities.

Standard Numbering Systems

The first output 1.1 is numbered 1.1 because it is the first of two outputs of Sub-Goal 1. Then because developing a lesson plan is the first activity of Output 1.1—call it Activity 1.1.1. The second activity of Output 1.1 will be Activity 1.1.2. This system of numbering will become important when you develop the budget and a schedule. Simply use the template and insert your activities using the example numbering system.

Download Example of Field Assignment 6 Log Frame 2 to see what this could look like or go to Text Box 6.2.

Box 6.2

Table 6.2 Course project example: Field Assignment 6 Log Frame 2

Sub-Goal	*Outcome*
(1) Sub-Goal 1 (Objective) [sub-components of the main goal, which when added together describe the main goal]	
Health and Hygiene Program	[Short-term positive outcome influencing mid-term positive outcome] 50 percent of the 300 children from 100 families in four communities will be free of chronic diarrhea through a six-month health and hygiene program
Output 1.1 [sub-components of the sub-goal, which when added together describe the sub-goal]	100 families in four communities participated in workshops on the benefits and practice of hand washing and received six months of follow up.
Activity 1.1.1	Develop workshop lesson plan; collect/develop workshop materials
Activity 1.1.2	Arrange workshop meetings with communities
Activity 1.1.3	Present workshops to 100 families
Activity 1.1.4	Provide six months follow-up

Activity 5: Monitoring and Evaluation: Indicators and Means of Verification

In this activity, you are going to begin developing a very simple monitoring and evaluation (M&E) plan. Your organization may have need of more detailed indicators and means of verification. Please feel free to follow your organization's guidelines.

Why a monitoring and evaluation plan? You want to be able to observe if your activities and outputs are working. If you are not achieving the results you want in the process of the project—and you can identify what isn't working—you can make adjustments to improve your results. An M&E plan also means that you can provide factual final reports, and learn from the results of your project.

Download the document called Example of Field Assignment 6 Log Frame 3. You'll see there are a number of previously empty cells that now have more information in them. Use this as a guide to fill in the empty cells in your project's log frame from last time. You can see what this looks like in Text Box 6.3.

Box 6.3

Table 6.3 Course project example: Field Assignment 6 Log Frame 3

(1) Sub-Goal 1 (Objective) [sub-components of the main goal, which when added together describe the main goal]			*[Long-term positive outcome] Impact*	*100 families in the western highlands of Guatemala have risen out of the cycle of poverty and lead healthy, prosperous, productive lives*
Health and Hygiene Program	[Short-term positive outcome influencing mid-term positive outcome] 50% of 300 children from 100 families in four communities will be free of chronic diarrhea through a six-month health and hygiene program		Outcome [Mid-term positive outcome leading to long-term positive impact]	100 rural families adopt beneficial health and hygiene practices into their lives, allowing children to grow and develop properly, be able to participate in education, and be prosperous, productive members of their communities
Output 1.1 [sub-components of the sub-goal, which when added together describe the sub-goal]	100 families in four communities participated in workshops on the benefits and practice of hand washing and received six months of follow-up.		100 families learn the health benefits and techniques of hand washing and adopt it for six months	Copy of sign-in sheets and photos of the workshops; copy of signed visit sheets, field staff notes, schedule and photos
	Activity 1.1.1	Develop workshop lesson plan; collect/ develop workshop materials	Lesson plan prepared, workshop materials collected	Printed copy of lesson plan, delivery or purchase receipts of materials, or photos

Activity 1.1.2	Arrange workshop meetings with communities	Prospective attendees filled in sign-up sheets for workshops and schedules prepared	Copy of sign-up sheets; copy of schedule
Activity 1.1.3	Present workshops to 100 families	Members of 100 families attend workshops on hand washing	Copy of sign-in sheets and photos of the workshops
Activity 1.1.4	Provide six months follow-up	Families visited once a month for six months after workshops	Copy of signed visit sheets, field staff notes, schedule and photos

Measuring Your Successes

Your activities and outputs need to be very clear so that you will know absolutely if you've achieved them or not, and they need to be measurable. In this activity you will add indicators that let you know that you have completed the activity, and a means of verification that creates certainty. For example, if you hold a workshop for 100 people—an indicator would be that they did indeed attend.

However, you need to verify that they did come—so means of verification could be the workshop's sign-in sheets and photographs of the workshop. Look at the example for ideas. Insert indicators and means of verifications adjacent to your outputs and activities as shown.

Activity 6: Outcome and Impact Statements

Why are you planning for impact? Development is not keeping pace with growing need. Part of that is because projects have not traditionally been designed to achieve long-term impact.

You began the process of designing long-term sustainability and impact into your projects by making sure that you had community buy-in. Next you investigated if the activities you had chosen had an evidence basis for solving the challenges that you identified. The next step in designing impact into the project is to make sure that you have a clear definition of the mid-term and long-term impact that you are trying to achieve: the ultimate goals.

Why Our Focus on Impact?

Escalating energy and food prices, shifting weather patterns, and increasing population pressures have led donors and NGOs to realize that we must research what has impact in development, and incorporate successful strategies into projects—and act with urgency. We need to shift from activity and output-based development to outcome and impact-based development.

In this activity you're going to write an impact statement that will be for the very long-term project goal that you're hoping to achieve, and outcome statements for each one of your sub-goals which represent mid-term achievements that will let you know that you are progressing towards the desired long-term impact. Your organization may have need of more detailed outcome and impact statements. Please feel free to follow your organization's guidelines.

Box 6.4

What is Impact? What are Outcomes?

According to Wilson-Grau's definition (2008), "Impact: Long-term, sustainable changes in the conditions of people and the state of the environment that structurally reduce poverty, improve human well-being and protect and conserve natural resources."

Outcomes are behavioral changes in partners—changes that take place over a three to five-year period that contribute to the long-term sustainable improvement in people's lives.

Two important points to note are:

1 Projects can only influence communities in making positive behavioral changes. For example, we can introduce the concept of hand washing to community members, but it is ultimately their decision to incorporate it into their daily lives. We cannot force them; we can only influence their decision.
2 You'll see that the outcome and impact statements are over to the right side of the log frame matrix. They are not part of the 'mechanical' part of the matrix where activities lead to fulfilling outputs, and outputs lead to filling sub-goals. Since we can only hope to influence our community, the outcomes and impacts are happening on another level that depends on sustained behavioral change. How do you ensure that communities will adopt these positive behavioral changes? Through community buy-in; through the community's sense of ownership of the project and its desired outcomes.

Impact Statement

If you return to the negative impact statements in the problem statement from Field Assignment 1, you will see that the example said: "These challenges contribute to stunting and restrict the ability of children to attend and concentrate in school." Where will that lead those children to be 15 or 20 years from now? Their lack of development and education will keep them trapped in the cycle of poverty and they may continue to be under-nourished and unhealthy.

Much as your goal statement was a mirror image or a positive version of your problem statement, the project's impact statement will be a positive version of the long-term negative impact. So the example impact statement says that "100 families in the western highlands of Guatemala have risen out of the cycle of poverty and lead healthy, prosperous, productive lives." That sums up the purpose of this example project: so these children will not grow up stunted, unhealthy, and poor, but will be well-developed, healthy, productive, and prosperous. Add an impact statement to your log frame. You will find an example in Example of Field Assignment 6 Log Frame 3.

Outcome Statements

In the same way as your outputs and sub-goals added up to equal your main project goal, your outcomes need to add up to equal your impact statement. The impact statement reflects the positive results of the project outcomes.

Hopefully, your project has influenced people's decisions to make changes in their behavior—such as incorporating hand washing into their daily routines. These changes are not something tangible that you can touch and feel (like a new school building) but are behavioral changes that result in healthier lives and increased well-being. These behavioral changes over three years or five years will create impact. The example outputs and sub-goals discuss people attending workshops and having improved health in the short term. But will they continue to maintain that level of health after your team has left? Outcomes describe the fact that people have adopted these healthy practices permanently into their lives.

The health and hygiene outcome for the example project states: "100 rural families adopt beneficial health and hygiene practices into their lives allowing children to grow and develop properly, be able to participate in education, and become prosperous, productive members of their communities." This positive outcome is a reflection of the corresponding section of the problem statement that is the basis for the project's health and hygiene program:

> 300 children in 100 families are frequently ill with chronic diarrhea caused by little knowledge of health and hygiene which contributes to stunting and restricts the ability of children to attend and concentrate in school.

Add outcome statements adjacent to each of your Sub-Goals. Download Example of Field Assignment 6 Log Frame 3 or go to Text Box 6.3 to see what this could look like. The download includes the full log frame for the course example project, including the agricultural income generation and climate-smart agricultural components.

Field Assignment 6 Step 2: The Detailed Project Budget

This activity begins a very important new step. You will start developing a series of parallel documents. You have now completed your log frame with all of your activities arranged in a numbered sequence. Over the next two steps you will transfer this information into a detailed budget and a project schedule.

The reason why it is important to have parallel documents is that different people in your organization, working on different aspects of the project, will each be using project documents that contain different types of information. It is important that they are all working with parallel information—and that they are all therefore working on the same project. The budget and schedule need to be an exact representation of the log frame. Having a detailed budget as shown in this activity's example allows you to make sure that you are going to be able to financially manage your project, and be able to report your financial management capabilities to your donors.

Activity 1: Creating the Budget

There are three steps in creating the budget:

1 copying and pasting the sub-goals, the outputs, and the activities into a spreadsheet;
2 expanding upon the activities with sub-activities;
3 assigning costs to each of the sub-activities.

Download the Example of Field Assignment 6 Budget. Open up your log frame. Copy from the log frame document and paste into the budget document your sub-goals, outputs, and activities right over the top of the example budget into the appropriate cells.

Be careful! The budget is a spreadsheet. Work freely in columns A through E; there are no hidden formulas there. Columns F, G, and H have simple formulas in the individual cells that will automatically total your results. You don't need to touch these for the purposes of this assignment. If you are not familiar with spreadsheets, I would leave those columns alone: they will automatically be filled in with the correct numbers.

Activity 2: Expanding upon the Activities with Discrete Sub-Activities

You will notice that under each of the activities, sub-activities have been added to the activities. The different sub-activities contain a variety of activities that need to be performed by different people, items that need to be purchased, and overhead expenses. Modify these line items to meet your project and organizational needs.

Activity 3: Assigning Costs to Each of the Discrete Sub-Activities

1 Under the unit column you can determine if the sub-activity should be expressed in terms of a month (example, a salary), or units (example, number of notebooks).
2 Under the quantity column you can select how many months of salary will be required or how many units you're planning to purchase.
3 Under the unit price insert your local cost for the items. Please note the currency you are using at the top of the log frame. You may be using $US or Kenyan Shillings—and your donor will need to know which it is.

Some donors may want your organization to share in the cost of the project. Therefore, two cost sharing columns are provided: one for the donor and one for your organization. In this example, the NGO is providing a 33 percent cost sharing. Enter units, costs, and cost sharing for your project into the budget. There will be many line items and cost categories that will be the same from activity to activity and you can simply copy and paste them throughout your spreadsheet.

You are done! If you are unfamiliar with spreadsheets, have your bookkeeper take 5 minutes to see that rows and columns add up correctly.

Additional Budget Design Information

You'll notice at the bottom of column A (below the end of the budget) that different budget categories (sub-activities) have been assigned number codes (1, 2, 3 . . .) to differentiate them:

Code	Categories
1	Director
2	Field Staff
3	Assistant

Create your own list of categories and number them. Then, enter the numbers into column A, adjacent to the appropriate category.

Next, when you feel your budget is complete, copy sheet one (Detailed Budget) into a new sheet two; change the name to Budget by Category. Then, sort the data in sheet two by column A. All of the number ones, twos and threes will be grouped together.

This can be very useful to let you see the totals of office expenses or travel, or to see how many months of time you allocated to field staff. For example, if you have a 12-month project, and in this sort process you added up the number of months for field staff and found that you have included only 9 months' staff time, you may have to go back to the budget and add more months. Or if you have 15 months of staff time, you may need to go back to the budget and reduce staff time in individual activities.

Conclusion

As you finish the budget, you will be able to see how much your project is going to cost. This exercise may show you that your project is getting too expensive and you may need to re-think some parts of your project. Donors frequently have cost/beneficiary limits. If your project is coming out at $100 per person, and your donor has a limit of $50 per person, then you might need to make adjustments. Find out from your colleagues what rule of thumb your organization uses for a cost/beneficiary figure, and adjust your budget accordingly.

Box 6.5

Table 6.4 Course project example: Field Assignment 6 Budget

Activity	*Detail*	*Unit*	*Quantity*	*Unit Price*	*Donor*	*NGO*	*Totals*
Sub-Goal 1	Health and Hygiene Program			$US			
Output 1.1	100 families participated in workshops on hand washing						
Activity 1.1.1	Develop workshop lesson plan						
1	Project director forms two-person program team	Month	0.5	800		400	400
2	Research best practices in hand-washing	Month	0.25	1200	300		300
2	Write lesson plan and how-to cards for workshops	Month	0.25	1200	300		300
2	Rehearse workshop with team members	Month	0.25	1200	300		300
5	Travel expenses	Unit	0	30		0	0
3	Assistant	Month	0.5	360		180	180
10	Office expenses	Unit	0.5	250	125		125

Activity 1.1.2	Arrange workshop meetings with communities						
2	Determine best time, location for the workshop	Month	0.25	1200	300		300
2	Develop a participant list	Month	0.25	1200	300		300
5	Travel expenses	Unit	5	30		150	150
3	Assistant	Month	0.25	360		90	90
10	Office expenses	Unit	0.25	250	62.5		63

Download Example of Field Assignment 6 Budget or go to Text Box 6.5 to see what this could look like. The full budget for the course project is in this download.

Field Assignment 6 Step 3: The Project Schedule

In this step you will transfer the log frame's sub-goal, output and activity information into a detailed project schedule. Having a parallel log frame, budget, and schedule will ensure that your bookkeeper, project manager, and field staff are all working on the same project—even though they may be focused on using different documents for their work. Consequently, you want the schedule to be an exact representation of the log frame. Having a detailed schedule as shown in this example also allows you to manage your project in a timely fashion, and be able to report your management capabilities to your donors.

Activity 1: Copying and Pasting Sub-Goals, Outputs, and Activities into a Spreadsheet

Download Example of Field Assignment 6 Schedule or go to Text Box 6.5 to see what this could look like. The full schedule for the course project can be seen in the download. Open up your log frame. There are two steps in creating the schedule:

1 Copy and paste from the log frame into the schedule your sub-goals, outputs, and activities right over the top of the example schedule.
2 Draw a bar chart. Adjust the blue squares to meet the timeframe requirements for your project.

The schedule can be very helpful in fine-tuning your budget. For example, after analyzing the time required for each activity, you may find the project will take 15 months—when you have only budgeted for 12. It is much easier to make adjustments now than suffer the consequences later.

Download Example of Field Assignment 6 schedule, or go to Table 6.5 to see what this could look like. The full schedule for the course project can be seen in the download.

Table 6.5 Course project example: Field Assignment 6 Project Schedule

Activity		*January*	*February*	*March*	*April*	*May*	*June*
SG 1	Health and Hygiene Program						
OPut 1.1	100 families participated in workshops on hand washing						
A 1.1.1	Develop workshop lesson plan	X					
A 1.1.2	Arrange workshop meetings with communities	X					
A 1.1.3	Present workshops		X				
A 1.1.4	Provide six months follow-up			X	X	X	X

Field Assignment 6 Step 4: The Compelling Two-Page Fact Sheet

Donors are busy people and have a dozen proposals sitting on their desks each day waiting to be read. As enthusiastic as you may be about a lengthy proposal you have written, handing it in this format to a donor may not be the best way to start off your first meeting: it will just look like more work to them. Anyway, if a donor is interested in your project, they will usually have their own proprietary proposal forms that you will need to fill out; their forms will typically become your final submission.

A good alternative is to hand them a 1½- or two-page clearly organized document: a fact sheet. They can scan it for 30 seconds or a minute, and quickly get a good understanding of your project. Properly done, they will also get a sense of your organizational capability.

This document needs to cover the nuts and the bolts of the project—and it needs to be compelling. The needs statement needs to be negatively compelling and your proposed solution needs to be positively compelling. Each one of the fact sheet's sections is only a paragraph long, but they all need to carry within them a compelling storyline.

A compelling story paints a picture that makes the reader feel "I was there." It can be a heart-wrenching story about an event in the day of a family suffering extreme poverty, or it can be a heart-warming story illustrating something wonderful that happened to a family as a result of your organization's work. The best compelling stories illustrate a single, human-centered image that supports the theme of your work and is interwoven with human emotion: something readers can relate to with a sense of urgency and immediacy—through joy or sorrow.

It is the human side of the project you have worked so hard on in these assignments to systematize. It is the hook that pulls at a donor's heart strings. It is the soft, intangible part of the project. It is what your impact statement is written about. It is why you are in development.

In the field, the story collector's job is to find a compelling story. Your first and second community assessments (Chapter 1 and Chapter 2) were ideal opportunities for compelling stories, and if you look through your notes and photos, you will find that you already have a collection of positive and negative story lines that you can choose from to use in your fact sheet.

Here are some examples of positive compelling story lines:

- An illiterate father who hadn't let his son attend school is invited to an NGO-led consciousness-raising workshop on math. Afterwards, he confided that he hadn't known how useful math was, but now that he sees its daily usefulness, he will encourage his son to enroll in school.
- An illiterate family has their fifth-grade daughter read them stories at night after dinner—opening a window to a new world and expanding future opportunities for the family through the eyes of their literate daughter.

- Through small but consistent earnings from NGO-assisted sales of her textiles, a poor woman was able to increase her family income enough to allow her daughter to attend school. Now, 16 years later, the daughter is preparing for her legal bar exams.
- A woman learned about the importance of nutrition for her children's physical and cognitive development, and through micro-enterprise savings, was able to buy the small lot next door to her home to plant a healthy vegetable garden and to raise chickens to improve her children's daily nutrition.

Writing Your Fact Sheet

The two hardest things about writing are getting started and being too self-critical early on. So just start writing and don't worry about how it sounds. When you have a first draft down on paper, read back through it and fix the obvious spelling and grammar problems. Then put it down, take a one-day break, and revisit it when you can re-approach it with a fresh mind.

Your next task is to begin crafting the document so that it is well written and compelling. You could spend an hour a day over three days editing a document like this. If you get stuck, ask a friend who has editing skills to help you. If you can't do this in two pages—including two or three photos—it is too long. Ask a friend to help you edit it down to two pages. Enjoy the process of playing with ideas, moving some things around and embellishing others. When you are happy with the outcome, have someone else read it. Something that is clear as day to an author may not be clear to another reader. Another person's comments can be very valuable in helping you to get your message across.

Look back through your photos and find two or three that represent a person, a project activity or a project installation. The photos should have people in them—and they should be your best people shots: sharp, simple in composition and colorful. The photos should complement and illustrate the message that your story is conveying.

If your donors have their hearts warmed and feel that you captured the essence of their mission in your project design, you will have a greater likelihood of developing a partnership. They should also be impressed with this well-organized, professional two-page presentation and sense that you would be a good organization to work with.

Activity 1: The Compelling Two-Page Fact Sheet

Download the Example of Field Assignment 6 Fact Sheet. It has been organized with section headings in red that you will eventually erase. They are there to help you understand the purpose of each section and their progressive order in the document. Go through the document and change obvious things like project names and dates and donor names but keep the format of the document intact.

Next you will be able to copy and paste, and adapt information that you have already written from your different assignments. This will save you time, it will also make this document parallel to your back-up documentation. Finally, you will craft this fact sheet so that it is compelling and enjoyable to read. This will create a new, people-centered window into an otherwise highly structured project design.

Let's go through the sections of the two-page document now.

Compelling Need

You need a negative compelling story line in this paragraph. Gather your Assignment 1 problem statement, your notes from the needs assessments, your summary of the community's needs, the original negative impact statement, and your impressions, and craft a negative compelling needs paragraph. Look for any photos you may have taken.

The Appeal

Donors don't have time to waste, so get right to the point about what you are looking for. Draw upon your log frame and look at your goal statement, your positive impact statement, your outcomes, and your budget for this paragraph. You need a positive compelling story line in this section: a call to action! Go to your budget and insert only the value of the donor's contribution (the 67 percent as shown in the example) of the total budget in this paragraph.

About Your Organization

Be very careful here. Keep this short. Donors are bored to tears with country statistics, organizational charts, and organizational history. If they want it, they will ask for it later. They just want to quickly assess your organization's capabilities. Highlight a couple of successes, mention partners you may work with so they will feel that you have a good support network, and mention the fact that you have experience in specific areas that will be important for this project.

How You Plan to Solve This Problem

This is a simple copy and paste of your Chapter 5 project outline.

Measuring Success

You have two options here. If you can highlight verifiable success of projects that your organization has concluded, do that. If you are a new organization, or this project discipline is new to you, use the information from the summary paragraphs from your Chapter 5 evidence-based activity assignment.

Budget Narrative

The project cost comes from your detailed budget. The narrative parts come from your outputs and activities within the log frame. This narrative needs to present cut-and-dried information in a friendly, compelling, "act now" sort of a way. Point out that the donor's contribution is only a portion of the overall budget, indicate what your contribution will be.

The Conclusion

This simply summarizes the six sections above. It can include portions of your outcome statements and your goal statement from your log frame. Download the Example of Field Assignment 6 Fact Sheet to see what this could look like.

Field Assignment 6 Step 5: How Will You Organize a Donor Presentation?

In this activity you are going to collect your completed assignment examples and have them bound into a presentation book. You can use this presentation book in a meeting with a potential employer, in a meeting with your boss when you're looking for raise, or in a meeting with a donor when you hope to impress them with the quality of your work.

This presentation book clearly shows what your project is about, it shows how you have built sustainability and impact into the project, it shows how you're going to manage the project successfully, and it gives your audience a sense of your capabilities.

Activity 1: How Will You Organize Your Ideas for Presentation to a Donor?

Collect your assignments. Clean them up in your word-processing program and in your spreadsheet program (make sure they look good and there are no typos), and print them out carefully. Collect a few of your best photographs of the community that you worked with and print out a single sheet of card stock with four or six photos and a project title on it for the front cover. Arrange your assignments in order and have them inexpensively bound with a clear plastic cover and a spiral binding. This booklet will only be 18 or 20 pages long.

It would be good to include the following:

- photos of your community with a project title printed on card stock for the front cover;
- results from your two community needs assessments;
- simple project outline with your research on potential activities;
- research on scientific evidence for your activities;
- your log frame;
- your budget;
- your project schedule;
- your two-page fact sheet.

Whether this is a potential employer, a current boss, or a potential donor, a presentation book like this has frequently made my meetings successful. Sometimes people in meetings will ask if they can 'borrow' it; my experience is that I never see the book again. This is the purpose of the two-page fact sheet: give them that. A good trick if someone asks to 'borrow' the book is to apologize and tell them that you need it for a meeting later in the afternoon. If, on the other hand, it is a donor, and you think that the book might help with a donation—give them the book.

Activity 2: Sharing Your Project with a Donor

I hope that you have been discussing your project with your co-workers and community members to get ideas and feedback from them. In this step, meet with someone a bit more challenging to share your project with: a donor. If this is not possible, consider meeting with your boss—or your professor, if you are a student—for ideas and feedback.

This exercise has several valuable benefits. The first is to start thinking about who your potential donors might be and thinking in terms of promoting projects to donors. The second benefit is if you haven't met with a donor before, this will give you the opportunity to experience what a donor appointment is like. The third obvious benefit is to get a different kind of feedback than what you've been getting from your co-workers and community members.

A donor, or your boss, or your professor may all have a different perspective on development projects than your co-workers do. They also have the power to let you proceed or not. It is very valuable to understand what their perspective is so that you will better be able to sell your project to them when your concept is fully developed. Receiving donor feedback and guidance at this point will help you improve your project so that it will be better received when you actually do submit it.

A donor will also be impressed that you would seek out guidance from them. It will plant a seed in their mind that you could be someone good to partner with on a project. Much in the same way as having worked to get a buy-in from your community members, you are now beginning to work to get a buy-in from the people who may fund the project. If they like what they see as the project design evolves, and truly feel that you have listened to their ideas and comments, they are going to feel a sense of ownership for the project. This will increase your chances of getting a project funded.

Identifying Potential NGO Partners and Donors that You Can Share This Working Project Proposal with

In this activity you will need to identify two international non-governmental organizations (INGOs) and two donors that you could partner with—and then make an appointment with one of them. How do you find donors? The first thing that I would do in my donor search would be to ask colleagues for suggestions. It is very likely that the organization you work for has a donor that they could introduce you to, who would be receptive to a meeting. Here are examples of other ways to identify potential donors:

- governmental offices;
- international embassies;
- international offices of cooperation agencies (DFID, USAID, the EU);
- local foundations;
- international foundations (Ford, Kellogg, Gates);
- individuals who make private donations (local and international);
- international NGOs looking for local partners to help implement projects.

Ongoing identification should be carried out continuously. Use the following to find possible donors:

- newspaper articles;
- e-mail news alerts for your country;
- networking with colleagues;
- RFPs (requests for proposals);
- get on the invitee list to formal receptions (of INGOs, embassies);
- your country's governmental offices;
- telephone directory;
- keep a database of this information with contact information, key words, specialization and grant size.

The Donor Meeting

Donors are busy people so you might need to make an appointment two weeks or a month in advance. When you first meet, you can explain that this is an initial project concept that you are hoping to receive feedback on before moving forward with concept development. The first thing that you should show the donor is your two-page compelling fact sheet that brings your project to life. They can scan this two-page sheet for 30 seconds and see what the project is about. As their interest grows, you can flip through your presentation book and show them the data that supports the project.

You may have met the donor before and have investigated the following information—but in case you haven't—in the meeting they will tell you one of two things:

1 The project concept falls within their goals and that you have some good ideas. Ask them for suggestions on what you could do to improve the project, based upon their experience, and how you can modify the project to best fit their programming guidelines. Find out if they have any submission deadlines and what levels of funding they provide ($5,000, $50,000, or $500,000?). Ask them if they have a rule of thumb for a cost/beneficiary budget figure. Ask them that if you develop the project further and incorporate their suggestions, can you return in two weeks to show them the result?
2 OR . . . The project concept is very interesting, but unfortunately it doesn't fit within their programming goals. At this point ask them: "What kinds of projects do fit within your programming goals?" Get very specific and take notes. They might not have anything that fits your organization, but it is good to find out what they focus on for future reference. Ask them if they can refer you to anyone at another funding agency that might be a better fit for your project.

However, they might have interest in an area that fits one of your organizational capabilities. Maybe you presented a health project to them, and you discover they are more focused on agriculture. Maybe your organization also does agricultural projects. Think quickly and say that you have a similar scale project in agriculture, can you return in a week to show it to them?

In either one of these cases, you have learned valuable information, and you have learned how to begin partnering with donors. If you are unable to receive funding from this first donor, polish up your presentation materials based upon the donor's input. Share your improved presentation package with your colleagues at your NGO, and ask if they can recommend other donors that you could approach.

Bear in mind that each donor has their own unique mission and their own proposal format that they will expect you to use. If you get a good lead for a donor from one of your colleagues, ask them about the proposal format that the donor requires. Frequently these formats are available online on the donor's website, or perhaps your colleague has an example proposal that you could use as a template. But don't give up! If you aren't successful with the first donor, or the second donor—you may well be successful with the third. And, in each meeting you will learn new information about how to better package your project. Good luck!

Chapter 6 Resources

Suggested Homework Assignment

The complete Field Assignment 6 homework to turn in will be:

1 a completed log frame for your project;
2 a completed budget for your project;
3 a completed schedule for your project;
4 a compelling, two-page fact sheet about your project;
5 a list of potential donors or NGOs that you could make an appointment with;
6 the name of the person that you made the appointment with;

7 your bound presentation book;
8 a brief description of the meeting with a donor and what types of suggestions they made.

Course Downloads

Go directly to this book's webpage, TimMagee.net/field-guide-to-cba/ to download the following resources:

- Example of Field Assignment 6 Log frame 1;
- Example of Field Assignment 6 Log frame 2;
- Example of Field Assignment 6 Log frame 3;
- Example of Field Assignment 6 Budget;
- Example of Field Assignment 6 Schedule;
- Example of Field Assignment 6 Fact sheet.

Part III

Sustainable Implementation

Part III of this field guide focuses on launching the project, implementing the project, including the community in project management, and leaving the community with a long-term management plan when they take the project over at grant's end.

Chapter 7 looks at increasing the community's engagement in the project before the project is launched by developing a community based management committee for one of the project components. This will ensure that they will learn by doing during implementation so that they will be able to continue with project activities after your NGO has left.

In Chapter 8, you will learn how to empower the community in preparation for receiving the project at takeover—and how to prepare them for long-term management of continuing program activities.

Community members may have things that they need to learn how to do. Some of them could be management and oversight techniques (for committee members); others could be practical skills and field techniques for carrying out project activities (for community members). In Chapter 8, you will learn how to assess capacity, and lead a capacity-building workshop.

Chapter 9 looks at establishing a participatory monitoring and evaluation plan for the community to use for the long term. This will help them spot when something is going wrong—15 years from now—and help them determine how to fix it.

7 Launch

Partnering with the Community

Leadership and Sustainability: Community Teambuilding

Chapter 7 quick-starts the section on project implementation. In this chapter you will plan and organize a workshop to develop a community based program planning and oversight committee—the community team that you will partner with during the project's launch, implementation, and handover. Examples of programs for which you might develop a committee could be for water management, disaster preparedness, flood control, soil restoration, forest management, agriculture, or alternative livelihoods.

The management of a project whose outcomes are projected in terms of decades needs to be carefully planned. If you play too large a management role, it will make it more difficult for the community to take over when you leave—therefore they need to be engaged in project management too. You need to make sure that during the community's involvement in the implementation of the project, they learn how to manage and maintain the infrastructure that was built, oversee activities that will continue into the future, steward resources, and support behavioral changes that were adopted.

Now is the time to start a management committee—prior to the launch of project activities. This will begin the process of bringing decision-making down to the community level as they participate in project start-up and implementation. They will learn new decision-making skills that they will need after you have gone. Community committees can be established with different levels of sophistication—based upon the complexity of the project and the capabilities of community members. A committee can be as simple as electing a group of community members to oversee community participation in project activities—or as complex as setting up a formal committee with bylaws, and state registration.

Many adaptation projects revolve around water. Let's analyze why you would want a village water management committee. If an NGO arrives in the community with funding to develop a water system, spends a year designing and installing the water system, and then leaves, who will oversee and maintain it long into the future? Without involving the community in a design appropriate for the village, and training them in maintaining the system, there may be as high as a 50 percent chance of failure of the system within two years. On the other hand, when a community is engaged in developing a system that meets their needs, is involved in project implementation, is trained in the system's maintenance—and when this process is overseen by a responsible committee—the system has a much greater likelihood of continuing to work over the long term.

Chapter 7 will help you accomplish two steps in the development of your project. You will do the following:

- write a teambuilding workshop plan for developing a community management committee;
- facilitate the formation of a community based project management committee.

What you'll need:

- computer and e-mail;
- workshop meeting with community members.

Timeframe:

- 5 hours:
  - 2 hours to develop a workshop plan;
  - 3-hour workshop.

Field Assignment 7 Step 1: Preparing for a Teambuilding Workshop

Activity 1: Researching Background Materials

You will need to make a preliminary decision about how the committee should be structured, based upon the complexity of the project that your committee will oversee and the level of education and experience that your community members have.

Perform a quick Internet search to find resources about developing community committees focused on one of the specialized components in your project. This project component should be one that will continue into the future and will need long-term oversight and maintenance. Examples could be a community water management committee, a community forest management committee, or a community health and hygiene committee. A lot has been done with forest and water management committees; if you're having trouble finding something on your program's specific field, you can research forest and water committee materials and adapt them to your project. Links are available in Chapter 7 Resources.

Activity 2: Organizing the Workshop

Discuss with your community representatives the idea of developing a project oversight committee made up entirely of community members. Explain how they will have the responsibility of overseeing the community's participation in the launch, implementation, and ultimate takeover of their project. Introduce the idea of setting up this committee in a short participatory workshop. Ask them to schedule a workshop, and to put together a list of people who should attend. They should be sure that the list includes representatives of different subgroups within the community—especially women and representatives of marginalized groups.

Begin organizing the workshop itself more than two weeks in advance. Make sure that you have all of your materials together—like large sheets of paper, and pens and markers. As this is a 3-hour workshop you may also need to plan snacks and drinks. My suggestion would be to have two to three colleagues accompany you to help with the workshop. Be sure and put someone who knows how to take photographs in charge of documenting the workshop.

Activity 3: Compile a List of Project Activities

The first assignment for your newly formed committee will be to select program activities that they would like to participate in during project implementation. Begin by looking at your log frame, budget, and schedule for activity ideas that would be appropriate for the community to be involved in—and activities which will continue into the future after your project cycle has ended. Then make up a list of the best activities to share with the new committee.

Bear in mind that the community will have time and financial constraints that need to be taken into consideration. However, by getting them involved in project activities, they will learn by doing and be better prepared to take over the continuing activities after your organization is done with the project. The list should be simple, but have enough variety that they can choose which activities are the most interesting for them.

Activity 4: Developing a Workshop Plan

A very simple version of a workshop lesson plan for developing a committee has been included at the end of this chapter in Text Box 7.2. Adapt this lesson plan to your project by using the information that you researched on the Internet about specialized committees. You can also look in Chapter 10 at other lesson plans to get ideas. A lesson plan will make the workshop much easier for you to facilitate. The plan should include the following key concepts:

1 consciousness-raising about the importance of having a committee;
2 going through a step-by-step procedure of forming a committee;
3 electing or appointing committee members;
4 asking the newly elected committee, for their first task, to choose project activities to participate in.

One suggested addition to the lesson plan might be to propose appointing an interim, volunteer committee who, over the course of six months, would develop the committee's structure, establish member selection criteria, ensure gender inclusiveness, set goals, and develop a 12-month plan. Let them know that you will be there to foster this process. At the end of the interim period, the interim committee could be elected for a second term—or they could be replaced by new members. This approach might make it easier during the workshop for participants to volunteer to be committee members if they realize it's simply for an interim basis.

Field Assignment 7 Step 2: Forming a Community Project Management Committee

Activity 1: Facilitating the Workshop

Make the workshop as participatory as possible by keeping the exchange of information as conversational as possible—this isn't meant to be a lecture. The first thing to do for your workshop is to make sure that you are well-prepared. That you have people to meet with, you have a place to hold a workshop, and you have the materials that you will need. Hopefully you have colleagues to help you facilitate the workshop. Your lesson plan will be very helpful in telling you what to do at each step so that you can stay on track. It would be good to begin the workshop with an icebreaker (sing a song or play a game). You want to make sure that everyone is comfortable being there, that they feel safe enough to say what's on their mind, and that there is a certain amount of enthusiasm in the room in order to encourage participation.

Make notes right onto the lesson plan to remind you about things that went well, things that didn't go well, and how long the different activities took. For example, you might find that the workshop only took two hours to do rather than the three hours you estimated. Large sheets of paper and markers are also a good way to take notes. As your community members say interesting things, you can jot them down on the newsprint and use them later as reminders of what was said and what people's reactions were. They will also allow you to draw organizational charts as visual aids if necessary.

I know that you will be tired at the end of the day but I would take a few minutes when you get home to go through the lesson plan and make a few more notes. The next time you give this workshop, you can clean up the lesson plan to reflect what really happened. I would also recommend meeting with your team the next day to get their perceptions of how the workshop went and to see if they saw any improvements that could be made the next time that you give it.

Be sure to take photographs. Put someone in charge of photos. Have them take:

- close-up detailed shots of participants;
- close-up detailed shots of the materials that you use;
- photos of interesting drawings that you might have been done on the newsprint;
- shots of the whole group;
- a few shots of you facilitating the workshop.

Go to Example of Field Assignment 7 in Text Box 7.1 to see what this could look like, or download the Example of Field Assignment 7.

Box 7.1

Course Project Example

Over the past few months I've been working with eight community contacts—two from each of the four villages that the project is centered around. Each one of the pair of village contacts came up with a list of people whom they would recommend to participate in a workshop on setting up a Community Agricultural Committee. This committee will help steer the implementation of the upcoming project which has two agricultural components:

1 an agricultural income generation program;
2 a climate-smart agricultural practices program.

We invited all of the people on their lists; 21 people attended. We discussed the importance of having this committee and had a brief discussion of the responsibilities that committee members would have. We also described a step-by-step process of how committees are formed, how committee members are chosen, and about the future oversight of activities.

I proposed the idea of initially setting up a six-month interim, volunteer committee who could organize the different facets of the committee—and then either be re-elected after six months or be replaced. Since they did not have experience in setting up nor running a committee, they felt it would be good idea to have six months of experience before they made a longer-term commitment.

We discussed how many people would be an ideal number to have on the committee, the roles they would play, and what qualities members should have. The participants felt that nine people would be a good number for the interim committee—two from each village—and a ninth person who would be a facilitator and would also provide an odd number so that the committee would not suffer from a tied vote. I also suggested that we should make sure that 50 percent of the eight committee members should be women and that the committee should also recognize marginalized groups within the community.

Seven of the workshop participants felt that there was too much responsibility in the committee and were hesitant to join. So we needed to choose nine people out of the 14 remaining. This turned out to be easier than we thought. Only six women attended the workshop and only a few people representative of marginalized groups attended—this quickly identified six of the nine possible members; and so we really only needed to choose two more people, plus the facilitator from the remaining eight.

There was one highly respected elder in the workshop that everyone felt would be a good facilitator, there was another individual who had strong leadership capabilities whom everyone thought would be a good committee president. And then there was one young man who had a university degree in agriculture; everyone felt that he would be a good member of the committee too. In the end no vote was necessary—and everyone seems satisfied with the result.

The committee decided to meet every two weeks prior to project launch—and invited the NGO's project manager to attend the meetings so that they would be able to learn more about developing a committee and setting goals. A date was set for the first meeting. Before the meeting adjourned, I provided the new committee with a list of project activities that I felt might be interesting to them, and asked them to select the activities that community members would like to participate in. I asked that they make this their first task at the first meeting. We all agreed that the workshop was a success.

Box 7.2

Course Project Example

Workshop Lesson Plan: Leading a Participatory Teambuilding Workshop: Forming a Community Agricultural Committee

- *Duration of workshop*: 3 hours.
- *Purpose*: What workshop participants will be able to do as a result of the lesson:
    - *Objective 1*: Participants will understand the importance of having a community management committee.
    - *Objective 2*: They will understand the step-by-step procedures for forming a committee and nominating members.
    - *Objective 3*: They will understand the importance of organizing the different facets of the committee.

Materials

- large sheets of newsprint and tape;
- colored markers.

Activity 1: Introduction to the Benefits of Forming a Committee. Duration 45 Minutes (Including a 15-Minute Ice-Breaker)

PURPOSE

Participants will understand the importance of having a community agricultural committee.

What to Do

1 Ice breaker: Introductions. Sing a song or play a game.
2 Introduction to workshop: Tell the participants what they'll be able to do as a result of the lesson.
3 Discuss the challenges that community members face with low agricultural incomes.
4 Discuss the challenges that community members face with reduced crop harvests.
5 Discuss how a committee can create a plan for creating market links that can increase income.
6 Discuss how a committee can provide direction, consistency, and a single point of contact for a buyer.
7 Discuss how a committee can help farmers sustain improved agricultural practices that can improve crop harvests.

Activity 2: Forming a Community Management Committee. Duration 60 Minutes

PURPOSE

They will understand the step-by-step procedures for forming a committee and nominating members.

WHAT TO DO

1 Discuss different ways that management committees can be formed.
2 Discuss the skills that members of the management committee will need to develop over time.
3 Suggest that an interim, volunteer committee could in six months develop the ultimate committee, set goals, and establish a plan.
4 Ask the group to suggest a maximum number of committee members.
5 Ask for volunteers from the group to join an interim committee.

Take a 15 minute break.

Activity 3: How to Get Started? Duration 20 Minutes

PURPOSE

Participants will learn how to develop an initial plan for getting started on this program.

WHAT TO DO

1. First, schedule the next committee meeting where the committee will begin the six-month organizational process.
2. Review project activities and make an activity list they would like to be in charge of.
3. Develop an outline of the next steps to take during the first six months. Suggestions for the next steps could be:
    a. to look at the strengths of the committee members;
    b. to seek training for six months in developing and managing a committee;
    c. to develop a six-month plan to develop the committee's structure, establish member selection criteria, ensure gender inclusiveness, and set goals;
    d. to develop a 12-month plan to include the formation of an agricultural income generation subcommittee or the formation of an improved agricultural practices subcommittee.
4. assign specific responsibilities to specific committee members.

Activity 4: Conclusion. Duration 30 Minutes

PURPOSE

To reinforce what has been learned.

WHAT TO DO

1. discuss and review what has been learned;
2. make sure that there are no more questions and that everyone understands what happened in this process.

Chapter 7 Resources

Suggested Homework Assignment

The complete Field Assignment 7 homework to turn in will be:

1. the name of the resource on forming committees that you discovered and a very short two- or three-sentence summary about the resource;
2. the lesson plan that you wrote;
3. the list of activities to present the committee;
4. a brief overview of the workshop, and what you felt the successes were.

Course Downloads

Go directly to this book's webpage, TimMagee.net/field-guide-to-cba/ to download the following resources.

- Example of Field Assignment 7.

Recommended Resources

Website addresses change frequently. Simply enter this book's webpage for current links to resources, or enter the author's name, the organization's name and the document's name into your web browser to find the most current link.

Bonita, M. and Payuan, E. *Village Forestry Handbook*, Ministry of Agriculture and Forestry Lao People's Democratic Republic. Available at: http://www.mekonginfo.org/assets/midocs/0003105-environment-village-forestry-handbook.pdf

FARM-Africa. *The Key Steps in Establishing Participatory Forest Management: A Field Manual to Guide Practitioners in Ethiopia*, FARM-Africa Ethiopia. Available at: http://www.farmafrica.org.uk/resources/Key%20Steps%20in%20Establishing%20Participatory%20Forest%20Management2.pdf

Stalker, L. *Why Some Village Water and Sanitation Committees Are Better than Others*, World Bank Water and Sanitation Program-South Asia. Available at: https://www.wsp.org/wsp/sites/wsp.org/files/publications/327200744509_saybetter.pdf

V&A Programme. Vulnerability and Adaptation Experiences from Rajasthan and Andhra Pradesh: Community Based Institutions, V&A Program. Available at: http://www.intercooperation.org.in/images/Climate%20Change%20-%20Case%20study%20on%20Community%20based%20Institutions.pdf

8 Sustainability

Capacity Building for Community Takeover

In this chapter, you will initiate the process of launching the project with the long-term goal of handing the project over to the community at the end of the project cycle.

Community Empowerment for Project Co-management

Your two years is up, your budget has been spent, and it's time to leave. Have you empowered the community to take full control at this point? In many development projects, as the project nears completion, community members have not been fully prepared to receive the continuation of project activities. You need to make sure that they have the skill sets they will need to implement and maintain the activities they will use in the ongoing adaptation programs.

In this chapter, you will begin a process of engaging project management committee members in the co-management of the project and in the implementation of activities. You will also learn to develop a workshop that will provide community members with the information that they need to implement continue ongoing activities. You will officially launch project activities by holding the first workshop designed to provide this capacity building. Finally, you will assemble a customized set of project management tools for the committee members to use during the co-management process.

Chapter 8 will help you accomplish five steps in the development of your project. You will:

1. engage the community management committee in project implementation;
2. empower the committee members to co-manage the project;
3. identify a program for the first community capacity building workshop;
4. write a workshop lesson plan;
5. launch project activities: facilitate the first capacity building workshop.

What you'll need:

- computer and e-mail;
- meeting with committee;
- capacity building workshop with community members.

Timeframe:

- 10 hours:
  - 1 hour for preparing management documents for the committee;
  - 2 hour-committee meeting;
  - 2 hours for workshop lesson plan and how-to card;
  - 5 hour-workshop.

Field Assignment 8 Step 1: Engaging Management Committee Members in Project Activities

Activity 1: Distributing Project Activities

The good thing about the community being involved in the implementation of a project is that their contribution will increase ownership, and the work that they do will be a learning opportunity for them in maintaining the outcomes of the project in the future.

In Chapter 7, you worked with the community to form a project management committee. During your last meeting you asked them to prepare a list of the project activities which community members would be interested in partnering in. Bear in mind, the community will have time and financial constraints that need to be taken into consideration when assigning tasks. In this first step you will organize these activities and formalize those commitments. You can see what this could look like in Text Box 8.1.

Box 8.1

Course Project Example

Activity 1: Distributing Project Activities

In Field Assignment 7, I worked with the four communities to form a Community Agricultural Committee to oversee the agricultural programs in our project: an agricultural income generation program, and a climate smart agricultural practices program.

I met with the new committee at their first committee meeting. They had gone through the activity list that I gave them two weeks ago. The committee decided to present the following activities as their contribution to the project:

1. They will organize the locations for all of the workshops on the correct dates and will provide tables and chairs as needed.
2. They will take responsibility for notifying community members about the time and the dates of the workshops.
3. For workshop activities which do have follow-up, they will take responsibility for notifying villagers about the follow-up calls and have someone to accompany us when we make the calls.
4. Each community will provide land for agricultural workshop demonstration plots.
5. Each community will provide a water source, fencing around the demonstration plots, and raw materials for making compost.
6. They will take responsibility for preparing the land in advance of the workshops (removing refuse or brush).

7 Each community will take custodial responsibility of the demonstration plots.
8 The committee also agreed that if the soil and water conservation component required a watershed restoration, they would be prepared to manage a tree seedling nursery and to provide labor for planting trees.
9 They will provide a single point person on the committee for us to communicate with in establishing workshop dates and locations.

As far as I'm concerned, this is wonderful. One of the complicated things about holding workshops is lining up the workshop locations and getting the word out. So they've taken on that responsibility which will save us a substantial amount of time—plus they're going to provide land for demonstrations and protect it and maintain it. They confided that they would like the demonstration plots to be adjacent to the village schools so that the schoolchildren can begin learning about gardening.

The other nice thing about the list they've developed is that I don't have to go through my log frame and divide up activities; it's very clear where the dividing line is between responsibilities. Finally, there is very little financial investment on the part of the community members in their activities. And, most of the activities that they're participating in will allow them to be engaged in the different steps of the project to see how the processes function and to learn how to do these things themselves.

I'm going to type this list up, and have the committee members sign it, I will sign it as well, and then we will each get a copy confirming their responsibilities. I will also attach a simplified log frame and schedule so that they can follow the project, and have a record of the commitment which my organization is making to the community.

Activity 2: Preparing Project Documents for the Committee

You've spent a fair amount of time during the assignments in this book developing your management tools: lesson plans, log frames, budgets, schedules, and fact sheets. These documents are reasonably sophisticated and might be overwhelming for your new partners. Perhaps they speak a different language than the documents are written in—perhaps they cannot read. It might be that some of the community members are better educated, but still you need to make sure that you have given appropriate tools to the people who will be co-managing the project. In this activity you are going to determine what management tools you should give them—and in what format—so that they will understand how the entire project is going to unfold.

Work on simplifying the presentation of the project so they will be able to understand it easily. If your community management committee is literate, you can simplify your log frame and give them that so that they can have a master plan of the project. Or you could give them the early project outline that you did in Chapter 5. If your community is not literate, you may need to do a series of simplified drawings that show the step-by-step process. Feel free to be creative—you want communication in a form that can be easily understood. Arrange a meeting with the committee to present them with their new tools and to discuss the project launch. Go to Text Box 8.2 to see what this could look like, or download the Example of Field Assignment 8.

Box 8.2

Course Project Example

Activity 3: Preparing Project Documents for the Community

Many of the committee members can read so I am going to give them a copy of my Chapter 5 project outline. This way, they can see what the entire project looks like:

Project Outline: Problem List Combined with Potential Interventions/ Activities/Solutions that I Chose

[Problem 1] Chronic diarrhea in small children

Health and Hygiene Program [Solution to underlying cause: Lack of knowledge of health and hygiene]

[Activity 1] Hand washing workshop and follow-up

[Activity 2] Point-of-use water purification system workshop and follow-up

[Problem 2] Chronic under-nutrition

Family garden and nutrition program [Solution to underlying causes: Lack of knowledge family nutrition; Overall shortage of food and specifically for the four months preceding the corn harvest]:

[Activity 1] Workshop and follow-up in family nutrition and home garden planning for nutrition

[Activity 2] Forming beds and planting seeds workshop and follow-up

[Problem 3] Insufficient income from agriculture

Agricultural Income Generation Program [Solution to underlying causes: Unpredictable dates for the start and end of rainy season, intermittent drought and erratic rainfall during the rainy season, flooding, and extreme weather events have reduced crop harvests and access to water]:

[Activity 1] Facilitate the organization of a community based farmer's association

[Activity 2] Survey local/regional businesses or markets that buy and sell agricultural produce in order to determine products they need on a routine basis

[Activity 3] Establish a market link and ask the businesses for their support in training programs and inputs for farmers to grow the products they need

[Activity 4] Using this input, launch training workshops on improved agricultural practices for these new crops and markets

PROGRAM RELATED TO CLIMATE CHANGE—BUT ALSO IN SUPPORT OF TRADITIONAL DEVELOPMENT CHALLENGES ABOVE

Climate-Smart Agricultural Practices Program [Solution to underlying causes: Unpredictable dates for the start and end of rainy season, intermittent drought and erratic rainfall during the rainy season, flooding, and extreme weather events have reduced crop harvests and access to water; Overall shortage of food and specifically for the four months preceding the corn harvest]:

[Activity 1] Identify expert specialist/extension agent in soil, water and agriculture to design and facilitate participatory workshops

[Activity 2] Participatory mapping and identification of local soil, water and crop challenges
[Activity 3] Consciousness-raising workshop on soil and water conservation and improved agricultural practices
[Activity 4] Farmer workshop on soil restoration and conservation techniques—including composting and mulching
[Activity 5] Farmer workshop on water conservation and management techniques—including water harvesting techniques
[Activity 6] Farmer workshop and follow-up on early maturing and/or drought-resistant crops for adapting to climate variability

Climate Change Awareness Program [Solution to underlying causes: A lack of knowledge of climate change and its near and long-term impacts]:
[Activity 1] Community based workshop and survey to identify their knowledge of climate change and its potential near and long-term challenges
[Activity 2] Consciousness raising workshops about climate change, its near and long-term impacts, and the need for a long-term adaptation plan
[Activity 3] Discussion and community prioritization of developing a long-term adaptation plan as a follow-up project

I am also going to give them a copy of my schedule—because it is very simple—and it will let them see the approximate dates of the upcoming workshops. I arranged a meeting for next week to present these tools and discuss the launching of the project.

Field Assignment 8 Step 2: Empowering Committee Members to Co-manage the Project

You've worked very hard up to this point to build on committee ownership. Until now, this has been largely theoretical: talking about activities that will happen in the future. In Step 2, you will begin the process of engaging community members in the co-management of the project and in the implementation of activities. This will be a capacity building experience for them in preparation for project takeover.

In the last step, you organized the activities that the community has committed to doing, and prepared a simplified presentation of the overall project. You now have the tools that will show them how their activities relate to the global picture of the project.

You are going to begin by developing a clear line of communication with the community committee. Simply meeting with committee members to share the new list of activities, the simplified project outline and the schedule is a good way to begin clear communication. You might discover in this first meeting that your presentation materials are too complex, and that you will need to present them in a different format.

Beginning to agree to do things in support of each other within certain time frames begins the process of co-management. One of the things the community has agreed to do is to organize the location of the workshops on the correct dates. The fact you will discuss this with them begins the communication process; the fact that they will follow through on this activity and that you will show up to facilitate the workshop begins the process of co-management.

The next step of the process is for you to observe which management skills they may be missing, and to ask for regular feedback in your meetings. You should begin making notes about things that they would like to learn about to improve their oversight of project responsibilities. Sharing these management skills with them is something that you can do by demonstration over the span of time, or you may choose to hold a specialized workshop in order to provide specific information (how to collect water fees, simplified bookkeeping, how to hold elections for future committee members). You might want to consider including someone from the committee to be part of your team so that they can see how you do what you do—so that they will be able to do it after you're gone.

Activity 1: A Meeting with the Committee to Share the New Management Tools

Part 1. Communication: Schedule a meeting to present your management tools to the committee—the list of activities, the simple project description, and the simple project schedule. Try to make your meeting as open as possible and where communication can be as participatory and two-way as possible. Strive to understand their feelings, their impressions, and try to get their honest feedback. Take notes.

Part 2. Co-management: The next thing to do in the meeting is to work with the committee to schedule a workshop on a project activity with community members for two weeks from now. Tell them what the first activity is about, discuss the date on the schedule, and ask if they would move forward with announcing the workshop to community members and finding a location. Fill them in on your plans, how long you feel the workshop will be, what time you propose to arrive, a little bit about what will be discussed at the workshop, and whether you propose providing snacks and drinks.

Part 3. Assessing management skills: I would pick a single page in your notebook that you can use over the next few months to add notes of possible management skills that could be developed or improved. The first step would be to ask the committee: what new skills would they like to learn or improve upon so that they can be an effective committee? Also, you should be observant and make note of things that you feel could be improved upon.

Part 4. Mentoring: Have a heart-to-heart discussion with yourself and think about whether it would be a good idea to ask one of the committee members to assist you in some of your project activities. This could be assisting you in a workshop, or shadowing you during follow-up activities. From a positive standpoint, this person will learn by doing how to run the project. The downside may be that you don't have enough time scheduled to do this extra bit of work—or maybe the community members don't either. Ask the committee members if they think it is a good opportunity as well; if so, ask them for suggestions of who would be the best person for you to work with. Go to Text Box 8.3 to see what this could look like.

Field Assignment 8 Step 3: Community Capacity Building: Adaptation Skill Set Workshops

This step will begin the process of developing a series of workshops for building the skill sets that community members will need to improve their capacity for adapting to climate change. These adaptation skill sets are the specific things that they will need to learn how to incorporate into daily routines.

Box 8.3

Course Project Example

Activity 1: A Meeting with the Committee to Share the New Management Tools

PART 1 COMMUNICATION

The list of activities that the committee gave me was pretty simple, and so after I organized it and showed it to them, they agreed that these items were still the ones that they felt committed to do.

I then showed them the simple project outline that I had printed out for them. Several of the men and one woman on the committee can read—but the others cannot. However, the outline was simple enough that they were able to describe to the non-readers what it said. Everyone was also in agreement that this was still the project that we had been discussing all this time.

The same with the schedule—the literate members of the committee were able to understand it—but it was a bit complex for the non-readers. One of the committee members had the good idea that across the top where it said January through December, they drew some little stick drawings indicating the types of things that happen in those months. For example, in the months May through October, they drew in little clouds with rain coming out of them. In November, they drew a picture of a farmer picking corn. And in December, January, and February, they drew pictures of farmers picking coffee. In March, they drew a picture of a farmer preparing his fields—and in April planting seed. With those stick pictures, the other committee members could understand what would be happening when. Everyone remained in agreement with the project and schedule.

PART 2 CO-MANAGEMENT

I've chosen to begin activities with the Climate-Smart Agricultural Practices Program. Activity 2 in this program is to do a participatory mapping and identification of local soil, water, and crop challenges. I proposed that it be held in two weeks—and that we break it up into two workshops so that the farmers, in the first workshop, could draw up a simple community map which would show their farming plots, forests, sources of water, and areas that suffer from extreme weather events.

In the second meeting, we would actually go into the farmers' fields with an agronomist to look at the soil and let the agronomist show them some simple field tests that they could do to help them understand moisture content and organic matter content. This information could be added to the map. We would also go to the locations that they put on the map that indicated the sources of water—so that we could get a better idea of the potential need for restoration or protection. We will also explore what the areas are like which suffer from a lack of water and from periodic storm damage.

Since the farmlands do not represent a very large area, we have realized that we could do these workshops in about four hours each. We decided to schedule each of the half-day workshops for two consecutive Saturday afternoons. I agreed that we would begin the workshops with a luncheon. It is going to be held in Cojol Juyú, Comalapa (about 48 miles from Guatemala City) at the village school—which is adjacent to the demonstration plot for the agricultural workshops.

PART 3 ASSESSING MANAGEMENT SKILLS

The committee members are a mixture of illiterate members and literate members of different educational levels. The educated members will be able to draft simple documents and do simple bookkeeping for the committee. My single biggest concern is that the educated members will not have a fully participatory committee because of their status and abilities. I didn't really want to bring this up in front of the full committee, so at an opportune moment I pulled aside the better educated members and expressed my concern.

They remembered how we had spent additional time making sure that the committee was inclusive of women and marginalized people, so they understood the reason for my concern. One woman had a suggestion: what if one of my staff members facilitated the meetings for the first few months, during which time we would also provide them with training on participatory facilitation? Everyone seemed happy with this, and we presented the idea to the full committee and gained consensus.

PART 4 MENTORING

Since the majority of what we are doing in our project is based upon a workshop format with follow-up, inviting a community member into our team is not an issue of time for us—so we are perfectly happy to do it. But I want to be careful of how I present the idea to the committee—because time is valuable for these villagers—and I didn't want to set up a situation where they felt obligated to do this even though they didn't really have the time.

We had a good discussion about it and they felt that the investment would be worth the effort—because otherwise how were they going to learn the background of managing this project? They asked if we would be willing to provide a stipend for the new team member to offset time that they would lose doing their other tasks. Since the workshops are only done periodically—I'm not hiring a full-time person—I just needed to come up with a stipend for a day here and a day there—and we certainly have the budget for that. Since the men were going to be workshop participants, the committee chose the educated woman who was on the committee to be our new team player.

For example, let's say that your farmers' soils have become depleted and your project includes a soil restoration program. One activity in your program is to begin increasing the organic material content in the soil both to improve soil quality for boosting crop production and to retain moisture in the soil. The skill set that you are trying to support is for farmers to understand the value of adding organic material to their soil—and the value of maintaining and continuing to add organic material over time. You are trying to encourage them to adopt sound practices and maintain them long into the future.

Box 8.4

Course Project Example

Activity 1: Identifying a Skill Set that the Community Needs to Develop

I looked at my log frame and my Climate-Smart Agricultural Practices Program and looked carefully at all of the activities. I chose "participatory mapping and identification of local soil, water, and crop challenges" as being the skill set that I want to use for this assignment in developing the workshop.

I feel that it is a foundational skill for them to learn for several reasons. One, it is important for them to see the entire complex picture of depleted soil, challenges with water, challenges with rainfall and challenges with crops—all exacerbated by a changing climate. I want them to be involved in this in a participatory fashion so that they will have ownership of the process.

I also want them to be part of the mapping process because I want them to see what a terrible state their resources are in before doing the upcoming workshops on soils, water and crops. In two or three years, their soil will be healthier again and they will have better control of their water. I want them to be able to remember back to how bad it was and make their own comparisons. I feel that this will help encourage them to adopt these practices permanently.

Activity 1: Identifying a Skill Set that the Community Needs to Develop

Look at your log frame and do the following:

1 pick an output that addresses a climate-related challenge;
2 pick an activity from that program that is a long-term, foundational element in your program.

My example for this assignment is the mapping workshop because it is part of one of my CBA activities. This is not a requirement for you to do a mapping exercise. If your workshop is on planting mangrove seedlings in an estuary—excellent. If your workshop is the first class of a series of vocational training classes—perfect!

Go to Text Box 8.4 to see what this could look like.

Field Assignment 8 Step 4: Two-Way Knowledge Transfer

Turning One of Your Project Activities into a Lesson Plan and a Take-home, How-to Card

The concept of sustainability also has implications for your organization. In working with NGOs I've discovered that many of them do not document their activities. This means that next year when they decide to do an activity again, they might not have names and addresses of partners they worked with or background information for the activity, or the specifics of how they conducted the activity. I try to encourage organizations simply to collect their loose documentation at the end of an activity and stick it in a folder. Even if it's not tidy or well organized, this simple task will protect 80 percent of the information they need for the next time.

Throughout this book you have been building a series of templates. You are designing a specific project and developing the documentation for that project. However, this documentation can be used over and over again for different projects and different activities just by making simple modifications to the original template. Be sure to save these examples of your work. They could even come in handy during a job interview or an interview with a donor about a new project. They will be examples of the quality of work that you are capable of doing.

In this step you are going to begin developing a lesson plan template—a lesson plan for a community capacity building workshop which will be different than the lesson plan for establishing a project committee. A well-written lesson plan can be used by your organization in scaling up activities in other communities with other staff members. Chapter 10 has a discussion on how to write lesson plans, and a collection of lesson plans on classic CBA activities.

This week is a process of projecting into the future what your activity will look like in a classroom setting. To write your lesson plan, you will need to visualize yourself presenting information to a group of people. You will need to use your imagination to think of fun ways for learning. When I wrote my first lesson plan, I sat down with a teacher friend and asked her to make suggestions on how to improve it. Her best suggestion was: "Don't forget to make the lesson fun."

Also in this step you're going to draw a 'how-to card' for your workshop participants to take home. The card should be an illustrated reminder of the different phases of the workshop. I try not to include words on cards because so many community members cannot read, and because in the countryside there are so many different languages. In the field, I will work with a local person to do the drawings because their drawings will be representative of the culture participating in the workshop.

Activity 1: Writing a Lesson Plan

I would like for you to find (in Chapter Resources or on the Internet) a manual that will give you sound how-to information about your workshop's main activity in order to help you develop a lesson plan.

In Chapter 10 you will find lesson plan examples of different types of activities. My suggestion would be to take one of these lesson plan examples which closely resembles your workshop and use it as a template. Start by writing a short outline of the activities that you will need to do in your workshop—and then paste those into the example lesson plan.

I first thought of the different steps that I would need to take to facilitate a participatory mapping workshop and then took one minute to write out a simple outline:

- introduce and explain the value of identifying and drawing a map of farm fields, terrain, crops, and water resources;
- draw a small-scale map;
- expand the small-scale map into a larger-scale, more exacting map;
- embellish the map with specific information;
- conclusion.

I pasted these line items into the lesson plan headings and then I was able to enter an activity or two beneath each heading. You will also need to use your imagination to think of how long each activity in the workshop will take—so that you can determine the length of the entire workshop. The only challenge you may face will be to dream up some fun activities for your participants to do. If you get stuck, ask a teacher friend if they can give you some ideas. If you would like to lead a participatory mapping workshop also, you can read the lesson plan in Chapter 10, Field Guide 5.

Activity 2: Drawing an Illustrated How-To Card

Take the step-by-step ideas from your lesson plan and think about what they would look like visually if you were to draw one illustration for each activity. Then decide how many illustrations you are going to need. Four to a page seem to fit well—so you could choose four illustrations or eight if it will be a two-sided how-to card. Example cards have been provided in Chapter 10.

Go to Text Box 8.5 to see what this could look like, or go to Chapter 10 to see other lesson plan examples.

Box 8.5

Course Project Example

Activity 1: Writing a Lesson Plan

I chose the lesson plan that I wrote several months ago on hazard mapping to use as my template because it had a similar layout as the upcoming workshop: it should be easy for me to copy and paste into this lesson plan.

Then I visualized what it would be like to lead a group of people through the process of mapping their village, their farms, their crops, and their water sources. I wrote a simple outline of activities that I would use, and then pasted them into the lesson plan. I also described the actual step-by-step sub-activities that we would be doing such as taping sheets of paper together, drawing, and adding details to the map.

I included the explanations of why we were doing these things. In order to make the workshop memorable, I needed to make sure that the participants were engaged in actually drawing the map. The lesson plan is for a half-day workshop and has five main activities, each with carefully described sub-activities. By including all of this information in the lesson plan, another staff member who wants to lead this workshop in the future will have everything they need.

Activity 2: Drawing an Illustrated How-To Card

I decided that for the illustrated how-to card that I should only include illustrations about the most important parts of the workshop. So I drew the map in stages. I first drew a very simple map with the shape of the village and the area of farm fields and the hills behind the farms. Then I duplicated that map and began adding roads and footpaths and streams. Then I duplicated that map and began adding houses and the shapes of the individual farm fields. I wound up with four drawings of the map in its step-by-step progression and I put two drawings on each side of one piece of paper. I left them black and white so that they would be inexpensive to photocopy.

If I were to do this drawing for a community workshop, I would first share these drawings with a few community members to get their input. I might have drawn something which isn't appropriate, or some of my images may not be easily recognizable to them. I would then try to find someone in the community who knows how to draw simple drawings, and have them prepare the final drawings. We are not looking for sophisticated illustrations; we are trying to transfer knowledge and want to make sure that our constituents will get value out of the how-to card.

Go to Chapter 10, Field Guide 5 to see what a mapping lesson plan and how-to card look like.

Field Assignment 8 Step 5: The First Community Skill Set Workshop

Congratulations! You are officially launching the implementation phase of your project. You are beginning to make their project real. This week you are going to hold the workshop that actually launches the activities of the project that you've worked so long to develop. Make the workshop as participatory as possible by keeping the exchange of information as conversational as possible.

Activity 1: Planning the Workshop and Touching Base with Your Point Person at the Community Committee

Begin organizing the workshop itself two weeks in advance. The first thing to do for your workshop is to make sure that you are well-prepared, that you have community members to meet with, and you have a place to hold a workshop. Check with your contact person to find out the best day for community members to attend a workshop—and to arrange the location for the workshop.

This would be a good time to also check and make sure that you have all of the supplies and materials that you are going to need—like large sheets of paper, and pens and markers for doing drawings and taking notes. If you are planning on providing a snack, make sure that that is well organized and that you will have colleagues to prepare and serve the snack so that you can fully concentrate on your participants.

Preparation

Since this will be your first mapping workshop in the community, it would be a good idea prior to the workshop to have a tour led by a knowledgeable villager. You may learn some things that you didn't know, you may have your memory refreshed about things that you did know once, but most importantly the information will be fresh in your mind for when you begin the workshop.

Activity 2: Facilitating the Workshop

Hopefully you brought some colleagues to help you facilitate the workshop. Your lesson plan will be very helpful in staying on track as it tells you what to do at each step of the workshop. It is good to begin the workshop with an icebreaker (sing a song or play a game) and introductions around the room. You want to make sure that everyone is comfortable being there, that they feel safe enough to say what is on their mind, and that there is a certain amount of enthusiasm in the room in order to encourage participation.

Take notes throughout the day right onto the lesson plan to remind you about things that went well, things that didn't go well, and how long the different segments took. For example, you might find that the workshop only took four hours to do rather than the six hours you estimated. I know that you will be tired at the end of the day but I would take a few minutes when you get home to go through the lesson plan and make a few more notes.

Be Sure to Take Photographs

Put someone in charge of photos. Have them take:

- close-up detailed shots of participants;
- close-up detailed shots of the materials that you use;
- photos of interesting drawings that you might have done on the newsprint;
- shots of the whole group;
- a few shots of you facilitating the workshop.

Go to Text Box 8.6 to see what this could look like.

Box 8.6

Course Project Example

Activity 1: Project Launch: Leading a Participatory Workshop in a CBA Long-term Skill Set

This workshop we are trying something new: we have invited one of the committee members to join our workshop team. This will be a great opportunity for them to learn the insides of launching and managing a project so that in two years when our NGO pulls out, they will know how to take over. On top of that, we discovered that she's a benefit for us too. She was able to alert us to cultural sensitivities that we should be aware of, offer translation of words or phrases that we were unfamiliar with—and she's a good troubleshooter when we need something done or fixed quickly. She is also able to forewarn us of potentially complicated people attending a workshop and of people who are in conflict with each other.

I also took part of a day to return to the village and tour the farm fields, forests, and water points with one of the committee members as my guide. I was fortunate in that I was able to see some of the farmers and talk to them for a few minutes, and to gain a much greater understanding of the challenges they were facing.

I have given a lot of workshops but I am really glad that I had my lesson plan. It helped me to get started and when little things came up that threw me off track, it helped get me back on track. I tried to be as inclusive as I could, in order to make the community part of the process and also so that I could collect new bits and pieces of information that I hadn't known before.

The workshop was about drawing a participatory community map especially focusing on farmlands and water sources. My main concern at the beginning of the workshop was would it become too disorganized with a number of people trying to draw one map. But it worked out fine, the participants were receptive, they contributed wonderful ideas and useful information, and seemed to understand what we were trying to do.

The resulting map is quite good and very colorful. We drew a small map first to begin understanding the spatial relationships between the different parts of the community. A couple of the men turned out to have good drawing skills and so they were elected to transfer the information from the small map to the much larger map.

We drew the map so that only the edge of the village was showing along the bottom, so we could see the roads coming out of the village leading to the farmlands, and we could see the main highway that travels along the edge of the village too. From there, the farm fields are spread out in three roughly adjacent areas that are partially separated by small areas of forest and encroaching hills. This was good because we could then see where the water sources came from in the hills—such as springs and streams.

Since I did not grow up in this village it was extremely helpful to me to see a bird's-eye view of how transportation routes (including the highway, small roads, and pathways), steep hillsides, level farm fields, water sources, and areas of hazard during extreme weather events related to each other.

We then took colored sheets of construction paper and decided on shapes that would represent additional features from the perspective of the farmers. I brought some removable masking tape so that we could stick these symbols to the map yet be able to move them around or take them down. Each farmer was allowed to indicate where his plot was. Consequently, the map is more artistic than it is exacting, but everyone seemed happy with the outcome. I feel like I have a good baseline of where the community is now that we can use for a comparison when the project finishes in two years. When the workshop was over, I photographed the map so that I could have a history of how it evolves over time.

All in all, everyone thoroughly enjoyed the workshop. There were many animated discussions and I think that everybody was very happy with the outcome of the map. They are very much looking forward to next week when I return with the agriculturalist to physically go look at their fields, soil, crops, water sources, forests, and flooding challenges.

I started the workshop at 1 o'clock, but still it took us until 6:00 to finish the map and have our wrap-up discussions. So, my 4-hour workshop actually took 5 hours. Partially because a lot of people had to stand around while the artists transferred the information from the initial draft map to the big map.

My colleagues and I took photographs and in the process of taking them we looked at them on the back of the camera to make sure that they were of good quality, they were in focus, and that we had a variety of different kinds of photographs.

I took notes on my lesson plan of things that were going well—and the things that I could improve the next time—and I also kept track of how long each of the exercises took for future reference. My helpers and our community team member were extraordinary; they were able to keep everybody engaged.

The next day the team and I got together and reviewed the workshop and made notes that would help us to do a better job the next time we give it. We should have done this in two workshops. We would recommend a relatively short evening workshop to do the initial activities and to draw the initial small draft map. The two artists could transfer the information to a larger map during the week, and then we could start the workshop again a week later to begin applying the more specialized information. We all agreed that the workshop had been a success.

Chapter 8 Resources

Suggested Homework Assignment

The complete Field Assignment 8 homework to turn in will be:

1 a list of project activities that the committee members would like to participate in;
2 an example of the simplified plan of the project to give to the committee;
3 the lesson plan and how-to card for your workshop;
4 a description of how the workshop went.

Course Downloads

Go directly to this book's webpage, TimMagee.net/field-guide-to-cba/ to download the following resources.

- Example of Field Assignment 8.

Recommended Resources

Website addresses change frequently. Simply enter this book's webpage for current links to resources, or enter the author's name, the organization's name and the document's name into your web browser to find the most current link.

CARE. *Climate Vulnerability and Capacity Analysis Handbook.* CARE. Available at: http://www.careclimatechange.org/files/adaptation/CARE_CVCAHandbook.pdf

Cash, D., Clark, W., Alcock, F., Dixon, N., Eckley, N., Guston, D. Jager, J., and Mitchell, B. *Knowledge Systems for Sustainable Development*, Proceedings of the National Academy of Sciences of the United States of America. Available at: http://www.pnas.org/content/100/14/8086.full.pdf+html?sid=7c343bfa-582c-425e-ac8a-84739494c330

FARM-Africa. *The Key Steps in Establishing Participatory Forest Management: A Field Manual to Guide Practitioners in Ethiopia*, FARM-Africa Ethiopia. Available at: http://www.farmafrica.org.uk/resources/Key%20Steps%20in%20Establishing%20Participatory%20Forest%20Management2.pdf

IFAD. *Good Practices in Participatory Mapping*, IFAD. Available at: http://www.ifad.org/pub/map/PM_web.pdf

IIED. *Participatory Learning and Action 54: Mapping for Change: Practice, Technologies and Communication*, IIED. Available at: http://www.planotes.org/pla_backissues/54.html

Regmi, B., Morcrette, A., Paudyal, A., Bastakoti, R., and Pradhan, S. *Participatory Tools and Techniques for Assessing Climate Change Impacts and Exploring Adaptation Options, Livelihoods and Forestry Programme*. Available at: http://www.lfp.org.np/publications.php?id=34

9 Impact

Milestones into the Future

Learning Tools: Participatory Monitoring and Evaluation

This chapter will help you design a participatory monitoring and evaluation plan (PME) so that community members can learn how to monitor and evaluate their project in real time during project implementation. This learning process will prepare them for taking over monitoring project outcomes after your organization has completed its work.

Monitoring and evaluation are used to determine if project activities are being properly implemented, are positively impacting project challenges, and are contributing to mid-term outcomes and long-term impact.

- *Monitoring*: Tracks project activities, outputs, and outcomes through record-keeping, observation, and surveys.
- *Evaluation*: Assesses monitoring data to see the change in results and tracks progress towards goals and outcomes. Evaluation also attempts to link a change in outputs and outcomes to project activities.

You will also look into the future and identify key milestones that the community needs to be aware of that will be indicators of the positive long-term progress of their project. This will help you develop a post-project monitoring and evaluation plan that will allow the community to continue to evaluate their project, learn from their project, and fine-tune activities in an effort to stay on course.

Why does the community need to plan for long-term monitoring? There are three reasons:

1. The community needs a way of evaluating if project activities, infrastructure, and resources are being managed and maintained properly by the teams in charge.
2. The community needs to be able to evaluate if their project activities are being successful at solving the challenges they face—or if the activities need to be fine-tuned over time to be more effective.
3. Monitoring will let them know if they have an impending problem where they will need to seek outside expertise for a solution.

This chapter will also help you develop a plan for finding future project expertise and also a long-term management plan for the committee to use in carrying on activities when your project cycle is over. You will then package these monitoring and management tools in a simplified format for presentation to the community's project committee.

Chapter 9 will help you accomplish five steps in the development of your project. You will do the following:

1 develop a participatory project monitoring and evaluation plan;
2 develop a community based post-project monitoring and evaluation plan;
3 develop a plan for finding future experts and resources when needed;
4 develop a simple, long-term project management plan for the community;
5 package and present these tools to the community management committee.

What you'll need:

- computer and e-mail;
- meeting with community management committee.

Timeframe:

- 10 hours:
    - 2 hours for participatory monitoring and evaluation plan;
    - 1 hour for community based post-project monitoring and evaluation plan;
    - 2 hour-committee meeting to assess potential current project expertise;
    - 1 hour for developing a plan for finding future expertise and resources;
    - 1 hour for developing a long-term management plan;
    - 2 hours for packaging a toolkit for a final committee presentation;
    - 1 hour-committee meeting for presentation of toolkit.

Field Assignment 9 Step 1: Participatory Project Monitoring and Evaluation

In Field Assignment 9, you are going to simplify your project's monitoring and evaluation plan so that the community can learn to monitor and evaluate project results and be able to carry the plan forward, post-project, into the future. This will help them to know they are continuing to meet the project outputs, outcomes, and long-term goals long after you and your organization are gone.

The first step is to reframe the indicators that you developed in your log frame so that they are readily identifiable by, and user-friendly for, community members; their PME plan will be a much simpler monitoring and evaluation (M&E) plan than the one in your log frame. In your meetings with committee members over the course of the two-year project you can continue to discuss these new indicators, involve them in monitoring activities, and discuss with them the successes and the lessons learned. At the end of two years they should be quite accustomed to their user-friendly M&E plan. This will also continue the process of bringing decision-making down to the community level. They will continue to learn new decision-making skills they will need when you hand over the project.

Activity 1: Developing a Baseline

The Chapter 8 participatory mapping exercise did a good job of providing a solid view of the community and the interrelationships between the village, farms, forests, water sources, and areas of hazard. Date and save this map for baseline information. You should then duplicate the map; this will be a new working map where new information can be applied to show project progress visually. However, the information currently on the map is not quantifiable. In order to create a baseline to compare with future monitoring data, you will need quantifiable data.

Your organizational monitoring and evaluation plan probably has more facets and greater detail than what the community needs for a participatory plan. Consequently, in this first activity, you are going to identify project indicators that will be relevant to community members—and the same ones that they should continue to monitor into the future. One simple way to do this is to go back to your problem statement and goal statement and look at the major challenges that the project is attempting to solve. These relate to the original challenges from the very first needs assessment, and are therefore meaningful to the community. Pick the ones that the adaptation components in your project are going to address. For example, the book's example adaptation component—Climate-Smart Agriculture—is working toward these goals:

- food security;
- increased agricultural income;
- improved agricultural production.

Several of the adaptation activities that support these goals include:

- farmers adopting soil restoration and conservation techniques;
- farmers adopting water conservation and management techniques;
- farmers using early maturing and/or drought resistant crops;
- watershed restoration and water management techniques.

These are project elements that all of the community members will be able to understand and relate to. You should collect baseline information on each of these different project elements. Propose that the committee members participate in the baseline surveys, and that over the course of the project they participate in surveys to collect midterm and end-of-project data to compare to this baseline information. By participating with you in monitoring and evaluation exercises during the project they will learn how to do it and will be able to continue after the project proper has ended.

For the book's example project, we will propose to the committee a simple participatory monitoring and evaluation plan using annual surveys: baseline surveys at the beginning of the project, midterm surveys at the end of year 1, and end-of-project surveys at the end of year 2. The surveys selected for community members to be involved in cover four project areas. Go to Text Box 9.1 to see what this could look like.

You will need to provide the committee with simple survey forms. Many organizations have developed rapid appraisal surveys for different development sectors that are less than one page long and only have a few questions. A quick Internet search can identify surveys that can apply to the specific activities in your project. In conducting the baseline surveys, you will learn much more detailed information than in the mapping exercise. Mid-term and end-of-project surveys will give the community information to compare to the baseline in order to evaluate project successes.

Your project may not be an agricultural project. Other examples indicators could include indicators such as these:

- the number of latrines built and used for one year;
- the number of water filters placed in homes and used for one year;
- the reduction in the number of children with chronic diarrhea;
- the number of family gardens installed and maintained for one year;
- the number of compost piles/bins built and used for one year;
- the amount of food harvested from family gardens in one year;
- the amount of money earned from selling excess vegetables from family gardens in one year;
- the change in family food security and nutrition after one year of having a family garden.

Box 9.1

Course Project Example

Activity 1 Developing a Baseline

For the participatory monitoring and evaluation two-year plan, we are using surveys to collect data.

FOOD SECURITY: ANNUAL SURVEY

- food security: percentage of families enjoying food security—and to what degree.

AGRICULTURAL INCOME AND AGRICULTURAL PRODUCTION: ANNUAL SURVEY

- agricultural income: income per farmer, per crop and size of field, per year;
- improved agricultural production: changes in yield: annual production per farmer, per crop and size of field, per year.

FARMERS USING IMPROVED AGRICULTURAL PRACTICES: ANNUAL SURVEY

- soil restoration and conservation techniques: number of farmers adopting and maintaining soil; restoration and conservation techniques in their farm fields;
- water conservation and management techniques: number of farmers adopting and maintaining water conservation and management techniques in their fields;
- agricultural soil restoration: condition of soil in farm fields based upon simple field tests;
- early maturing and/or drought-resistant crops: number of farmers adopting the use of early maturing and/or drought-resistant crops.

WATERSHED RESTORATION: ANNUAL SURVEY

- watershed restoration: percentage of tree cover in community watershed;
- water management:
    - availability and quality of domestic water at key times of year;
    - impact of watershed restoration and water management techniques on reducing flooding.

There are many excellent online resources specific to PME. Look for resources that might list ideas for indicators that are specific to your project.

Activity 2: Packaging the Participatory Monitoring and Evaluation Plan

The next time you attend a meeting with the committee, you should present this simple plan to them to get their feedback. Show them copies of the simple survey forms. They may have valuable information for improving the plan. Make sure the committee members understand the plan, the reasons for the surveys, and feel as if they will be able to learn this process over two years. You may need to make modifications or simplify the plan. When you feel that the committee is comfortable with the plan, type up a clean copy for presentation to them, and include copies of the surveys.

Field Assignment 9 Step 2: Milestones: Post-Project Monitoring and Evaluation

Two years from now, at the point of project handover, you are going to want to prepare for the committee a second, post-project monitoring and evaluation plan. This end-of-project plan will be looking at long-term outcomes and will allow the community to continue to evaluate their project, learn from their project, and fine tune activities in an effort to stay on course.

Activity 1: Identifying a Roadmap of Key Project Milestones

You need to look into the future and identify key milestones that will let the community know if they are on track in future decades: milestones that they can use to gauge the success of their project, or determine if there is a potential challenge looming ahead. These are to be tangible things that they can easily observe to make sure that they are fostering positive project outcomes and long-term impact.

Looking at their reforestation project—what are they trying to achieve in five years, or ten years? What are indicators that will let them know that things are going along well? In the reforestation project, have the trees have survived and grown? Has there been less damage from flooding? Has the spring behind the village become recharged with water? What are negative indicators that there may be a challenge looming ahead? How will they know that an insect infestation is a serious problem? At what stage during the development of a problem will they need to realize that it is serious and that they should contact an expert?

In this course example project, the key challenges and activities used in the Step 1 are the same project elements that they will want to continue monitoring post-project. Consequently, they can continue using the same survey-based plan from Step 1 to see if they are meeting the following goals and outcomes in the long term. Your project may be different and require shifting to monitoring different post-project outcomes. Go to Text Box 9.2 to see what future milestones could look like.

Field Assignment 9 Step 3: Finding Future Expertise and Resources to Address Future Challenges

Does the community have access to resources which they may need for repair, maintenance, or technical support in the future? What can you do before you leave this project to help the

Box 9.2

Course Project Example

Activity 1: Milestones of Success

- families have 12 months of food security;
- farmer income per unit of land has increased over baseline from the beginning of the project;
- crop production per unit of land has increased over baseline from the beginning of the project;
- farmers are continuing to use improved agricultural practices;
- families are less vulnerable to flooding; flooding in farm fields is greatly reduced;
- community members' access to clean water has increased over baseline from the beginning of the project.

This is a list of project milestones that will let the community know if they are on track. The committee will be able to use the same PME survey-based plan developed in Step 1 to determine if they are achieving these milestones. They can make the decision to conduct the surveys every second or every third year instead of annually.

The post-project participatory monitoring and evaluation plan is simply a continuation of the Step 1 plan.

FOOD SECURITY: ANNUAL SURVEY

- food security: percentage of families enjoying food security—and to what degree.

AGRICULTURAL INCOME AND AGRICULTURAL PRODUCTION: ANNUAL SURVEY

- agricultural income: income per farmer, per crop and size of field, per year;
- improved agricultural production: changes in yield: annual production per farmer, per crop and size of field, per year.

FARMERS USING IMPROVED AGRICULTURAL PRACTICES: ANNUAL SURVEY

- soil restoration and conservation techniques: number of farmers adopting and maintaining soil restoration and conservation techniques in their farm fields;
- water conservation and management techniques: number of farmers adopting and maintaining water conservation and management techniques in their fields;
- agricultural soil restoration: condition of soil in farm fields based upon simple field tests;
- early maturing and/or drought-resistant crops: number of farmers adopting the use of early maturing and/or drought-resistant crops.

WATERSHED RESTORATION: ANNUAL SURVEY

- watershed restoration: percentage of tree cover in community watershed;
- water management:
  - availability and quality of domestic water at key times of year;
  - impact of watershed restoration and water management techniques on reducing flooding.

community find resources when they run into a future problem? In the community forest example, it could be an insect invasion on a certain species of tree. This may not require financial resources to solve, but it may require technical assistance from a forester. Further, this may be a problem that arises 20 years in the future.

You need to leave them with information on how they will be able to find technical support on their own. From your standpoint this might seem very simple, but if they are a remote, relatively uneducated community, they might not know how to start looking. They will need simple information on how to do this. It is important for community members not to only learn about current institutions and organizations that they can contact, but it is also important for them to learn about how they can research new organizations in the future; contact information that you give them today may have changed 20 years from now.

Activity 1: Developing a Plan for Finding Future Technical and Financial Support Partners

Make an appointment for a 2-hour meeting with the community management committee to find out their knowledge of potential resources. Ask them if there have been other organizations that have worked in their community in the past. Ask them if they know of other organizations that are working in surrounding villages. Ask them if they have family members or friends who have connections with institutions such as universities. Ask them for the names and contact information for people in these institutions and organizations.

Make a list of these different organizations on a large sheet of paper in a column on the left. Next to each organization have the committee members discuss what types of services and support the organization could offer them. When they have exhausted their list of organizations, share with them other organizations that you know about that could potentially be of benefit, and what services or expertise these organizations offer. Add them to the bottom of the list.

After the meeting, revise the list so that it contains the most appropriate organizations for them to approach. Next, in case contacts change over 20 years, write out a quarter page numbered list of instructions of what they could do 20 years from now to find a new institution or a person that they could approach with their challenge. Go to Text Box 9.3 to see what this could look like.

Field Assignment 9 Step 4: Developing a Long-Term Project Management Plan

Your two years is up, your budget has been spent and it is time for your NGO to leave. Have you empowered the community to take full control at this point? At the end of the last

Box 9.3

Course Project Example

Activity 1: Developing a Plan for Finding Future Technical and Financial Support Partners

I arranged a meeting with the committee to go through this exercise. We identified seven current organizational partners that they could approach for assistance.

- INTECAP: Guatemalan Government/private sector partnership that provides vocational training;
- MAGA: Guatemalan Ministry of Agriculture;
- University of San Carlos Department of Agriculture;
- University of San Carlos Department of Forestry;
- INAB: National Forestry Institute;
- MARN: Guatemalan Ministry of the Environment and Natural Resources;
- SESAN: Guatemalan Secretariat of Food Security and Nutrition.

I spoke with the committee about the long-term nature of maintaining the program components. We discussed the possibility that in 20 years they might run into a problem where they need technical assistance. We also discussed the fact that the committee members might change, and contacts at the institutions might change. We listed ways that new committee members could start fresh and contact new people for technical assistance.

Here are a series of steps that future committee members could take to find new experts and resources:

1 Keep the original project documentation in a safe place for presentation to a new, potential contact.
2 Keep a list of the main program components that have been maintained by the community.
3 Photocopy a standard Guatemala map showing the location of the community.
4 Write a half-page summary describing the problem.
5 Take several photographs of the challenge they are experiencing. In the case of plant problems, provide a sample.
6 Determine if solving the challenge would require a visit by an expert to the community.
7 Ask village leaders from surrounding communities if they know of NGOs that are currently working in the communities.
8 Meet with a member of the NGO to see if they can provide technical assistance for the problem at hand.
9 Do any community members (or or their relatives) attend university or work at one of the institutions listed above?
10 Get someone with a good telephone presence to call the appropriate organizations listed above. Ask for contact information for specific people and get their telephone numbers.

11 Call these technical experts and very briefly describe the problem and ask if they could provide assistance.
12 Let contacts know that you would be available to travel to their office and bring photographs and samples.
13 Upon identifying the most appropriate person to provide technical assistance, develop a plan to work with them.
14 Determine if this would be a free extension service, or if there would be costs involved.

chapter you set up a project management committee to oversee, manage, and maintain the program components that would continue after you're gone. You need to make sure that the committee has a management plan that fits their capacity for the continuation of these elements of the project. The management plan will be very similar to the post-project monitoring and evaluation plan. Go to Text Box 9.4 to see what this could look like.

Box 9.4

Course Project Example

Activity 1: Developing a Committee Plan for Long-Term Management and Maintenance of the Project

The Step 2 participatory monitoring and evaluation (PME) plan identifies project elements that committee members will want to continue managing in an effort to maintain resilience to a changing climate. Having this PME plan simplifies the development of the management plan because they are essentially the same.

COMMUNITY AGRICULTURAL COMMITTEE LONG-TERM MANAGEMENT PLAN

These are the long-term adaptation outputs and outcomes the community hopes to maintain. The milestone indicators of success from Step 2 are:

- families have 12 months of food security;
- farmer income per unit of land has increased over baseline from the beginning of the project;
- crop production per unit of land has increased over baseline from the beginning of the project;
- farmers are continuing to use improved agricultural practices;
- families are less vulnerable to flooding; flooding in the village and farm fields is greatly reduced;
- community members' access to clean water has increased over baseline from the beginning of the project.

The Community Agricultural Committee needs to be responsible for the following actions as part of the management plan:

- The monitoring and evaluation plan and the long-term management plan should be a permanent part of committee activities.
- The committee shall monitor whether farmers and community members are continuing support activities and encourage them to reinstate them if they are not doing so:
  - encourage farmers to continue using soil conservation and restoration techniques;
  - encourage farmers to continue using water conservation and management techniques;
  - encourage farmers to continue using crops appropriate for the changing climate conditions;
  - encourage community members to be good stewards of the reforested watershed;
  - encourage community members to properly manage and maintain water management techniques and infrastructure;
  - encourage farmers to continue successful agricultural income activities.
- Monitoring data shall be routinely evaluated to determine if project outputs and outcomes are successful.
- If not, the committee should evaluate support activities to see if they are still effective or if they need to be fine-tuned.
- When a challenge arises, and the community needs outside support, the committee will need to take action and find expert support.

Since the outputs and outcomes in the PME plan are virtually identical to those in the management plan, it should be straightforward for the committee to continue oversight of project activities.

Field Assignment 9 Step 5: Packaging and Presenting the Post-Project Toolkit

Activity 1: Present the Final Report to the Community

Field Assignment 9 should be organized, cleaned up, and presented to the community in a format which will be understandable and useful to them. Set up a short, final meeting with the community management committee to review the compilation of the community based monitoring and evaluation plan, instructions for finding project expertise in the future, and the committee's management plan. Make sure that committee members are comfortable with the compilation and understand it. Ask for feedback in case other revisions need to be made.

At the end of your meeting you should have a discussion about project follow-up that your organization will provide. Your NGO may have only budgeted in your original project funds to return in one year to do a follow-up evaluation that would satisfy a donor—or you might have a more involved follow-up plan. But your follow-up plan will need to be discussed with the committee so that they will know exactly what they can expect. Go to Text Box 9.5 to see what this could look like.

Box 9.5

Course Project Example

Activity 1: Present the Final Report to the Community

I typed up a copy of the post-project monitoring and evaluation plan, the plan for seeking future expert support, and the long-term management plan, returned to the committee, and went through it carefully. I asked for feedback and questions throughout the presentation in order to make sure that the information was easily understood. They agreed that it was what we had been discussing and they were happy with the presentation.

Finally, I spoke with the committee about the fact that we had planned for follow-up meetings with them every three months for one year, and then a final project evaluation at the end of that year. Beyond that, we did not have funding to continue the follow-up. I suggested that I would contact them a few weeks before my quarterly follow-up meetings so that they could prepare a list of questions or challenges that they would like to discuss with me. Everyone was fine with this arrangement—the committee felt that the project was clearly defined, that there had been enough capacity building, and that one year of follow-up would be sufficient.

Chapter 9 Resources

Suggested Homework Assignment

The complete Field Assignment 9 homework to turn in will be:

1 outline of your baseline plan;
2 source and name of the rapid appraisal surveys your discovered;
3 outline of the participatory monitoring and evaluation plan to present to the committee;
4 list of potential future organizations and experts to approach in case of future challenges;
5 outline of your long-term management plan;
6 the packaged toolkit to submit to the committee.

Course Downloads

Go directly to this book's webpage, TimMagee.net/field-guide-to-cba/ to download the following resources.

- Example of Field Assignment 9.

Recommended Resources

Website addresses change frequently. Simply enter this book's webpage for current links to resources, or enter the author's name, the organization's name and the document's name into your web browser to find the most current link.

CARE. *Framework of Milestones and Indicators for Community Based Adaptation*. CARE. Available at: http://www.careclimatechange.org/files/toolkit/CBA_Framework.pdf

Jain, S. and Polman, W. *Training Module on Participatory Community Monitoring and Evaluation*, FAO. Available at: http://www.fao.org/docrep/006/AD346E/ad346e0e.htm

Mathie, A. and Foster, M. *Participatory Monitoring and Evaluation: A Manual for Village Organizers*, Coady International Institute, St. Francis Xavier University. Available at: http://coady.stfx.ca/resources/abcd/SEWA%20PME%20Manual.pdf.NGO

Programme Karnataka-Tamil Nadu. *Participatory Monitoring and Evaluation: Field Experiences, Intercooperation*. Available at: http://www.sswm.info/sites/default/files/reference_attachments/Intercooperation%202005%20Participatory%20Monitoring%20And%20Evaluation.pdf

Part IV

Tools and Field Guides

Part IV is a collection of hands-on field tools and activities for your use in your project. The information is compiled into field guides that come complete with how-to information, illustrated hand-outs, and workshop lesson plans. This information is designed to be accessible, useful, and understandable to people of different levels within your organization. The activities are simple enough for implementation by field staff with basic skills, and for adoption by community members with basic capacities for sustaining activities.

Ten examples of project activities have been chosen that address universal adaptation challenges that communities face in the areas of water, food security, agriculture, disaster risk reduction, and livelihood diversification. They very likely represent solution-based activities that you could use in designing and implementing your community's project.

10 Tools and Field Guides

Chapter 10 presents a series of field guides. A field guide consists of a short description of an activity that you will implement with your community, an illustrated how-to card of the activity to hand out to community members, and a lesson plan on how to lead a participatory workshop on the activity. The field guides can be used as-is or used as modifiable templates; you simply pick the field guide that most closely matches your activity and modify it to best fit the new activity and the context of your community.

The field guides chosen for this chapter represent ten of the most universal challenges that communities face in adapting to a changing climate. The ten workshops each represent what would be the very first workshop of a series that you would offer to a community. Taken together, that series of workshops would create a comprehensive program within your project, such as a health program or an agricultural program. These introductory, consciousness-raising workshops are meant to be reinforced and expanded upon with extension activities, and by follow-up that you provide as part of your project.

This specific set of field guides represent no-cost/low-cost activities. They will be easy for you to implement and for your community members to adopt. These accessible features mean that while you are seeking funding for your entire, comprehensive project, you can get started empowering your community today with this very powerful set of tools.

Chapter 10 will help you in the development of your project. You will do the following:

- select and adapt field activities for your project;
- learn to develop your own field guides, how-to cards, and capacity building workshop lesson plans.

Knowledge Transfer: Developing Field Guides and Lesson Plans

The following field guides are available:

The Community

Field Guide 10.1: Participatory Community Needs Assessments.
Field Guide 10.2: Participatory Capacity and Vulnerability Assessments.

Food Security

Field Guide 10.3: Preparing Family Garden Beds and Planting Seeds.
Field Guide 10.4: Soil Restoration and Conservation for Smallholder Farmers.

Water

Field Guide 10.5: Participatory Mapping of Soil and Water Resources.
Field Guide 10.6: Agricultural Soil and Water Management for Sloping Land.
Field Guide 10.7: Household Rooftop Rainwater Harvesting.
Field Guide 10.8: Community-Level Water Harvesting.

Community Based Disaster Risk Reduction

Field Guide 10.9: Overview of Developing a Community Based Disaster Risk Reduction Plan.

Diversifying Livelihoods for Increased Resilience

Field Guide 10.10: Diversifying Livelihoods through Market Links.

Knowledge Transfer: Workshops

The implementation of your project will take many forms. There may be an infrastructure component if your organization is building a bridge or a school building, but, frequently, projects begin with meetings, classroom instruction, field instruction, and workshops. This chapter uses the term workshop to mean a range of types of meetings. A workshop represents the interface between transferring ideas and knowledge between you and your community members. Starting a farmers' field school, facilitating a needs assessment, or teaching families how to grow a range of nutritional fruits and vegetables all fall within this chapter's definition of workshops.

There are three types of workshops in this chapter that represent three different stages of project development: (1) those designed to gather information about the community in a participatory manner; (2) those meant to present an overview of a topic to community members; and (3) those designed to be strictly for training community members to do a single activity. You can adapt these workshops to your project theme for use at the appropriate stage of your project.

Integral to these workshops is the concept of transferring knowledge. The knowledge to be transferred in community based adaptation needs to be a two-way transfer. You used this idea extensively in the initial community needs assessment and again in the vulnerability assessment.

In your project, you may choose to have an agricultural workshop to learn about your community's traditional agricultural strategies. A week later, you might return for another meeting to share with the community a few improved agricultural techniques that would support their traditional strategies. By having this two-way communication you can begin forming the process of building bridges across which knowledge can pass and action can begin.

A second element in knowledge transfer is that the exchange of information should ideally be between people who are culturally similar to each other. It could be challenging for a highly specialized university professor from Europe to enter a remote village of subsistence farmers in order to share scientifically-based information about modern agriculture. A better strategy perhaps, would be for the professor's knowledge to pass through several layers of people—becoming somewhat culturally filtered along the way—and to then be presented to the community members by someone who is more culturally akin to the community members than the university professor would be.

Experts and remote, rural community members have different positions and expectations. Bridging these positions and expectations requires mutual understanding—understanding that can be complicated by language, culture, and experience. Field staff who come from a culture similar to the community members can bridge a number of gaps including expectation, language, cultural acceptability, educational levels, trust, and capacity to adapt to new techniques.

Field Guides for Workshops

I start preparing for a workshop by developing a short, concise, one-page field guide that helps me in a number of ways. It allows me to compile how-to information from diverse sources into a guide that best fits my context. It also gives me the basis for a lesson plan—and a single-sheet handout with the field guide printed on one side and illustrations printed on the other. A three-quarter-page guide also limits how much I can accomplish in a workshop. It forces me to concentrate on a single specific activity that I can present in two or three hours. Examples of specific activities could include:

- the importance of hand washing and the correct way to wash one's hands;
- using point-of-use water filters;
- how to build a compost pile for a family garden;
- how to dig a garden bed and plant seeds.

Here are a few simple steps to develop a field guide for your workshop:

- clearly identify a discrete challenge that you're hoping to address;
- select an activity from your log frame that will address the challenge;
- search the Internet for manuals and handbooks describing how to implement the activity successfully;
- refine your search by seeking peer-reviewed papers to determine the evidence-based effectiveness of the different approaches found in the manuals;
- choose an activity that shows strong evidence of solving the challenge in your community's context;
- write a short one-page summary field guide on how to implement this discrete activity in a workshop;
- illustrate a how-to card to hand out to workshop attendees;
- copy and paste sections of your one-page field guide into a workshop lesson plan to conduct a community workshop.

What I do when I am producing a field guide is that I will read the available literature on an activity—peer-reviewed scientific documents as well as handbooks and manuals. I will then look at the adaptive capacity of my community as well as their traditional knowledge and I will begin to narrow down my choices of step-by-step workshop exercises and activities.

For example, let's look at a farmers' soil and water conservation program. That would be rather difficult to present in a half-day workshop. However, breaking it down into discrete activities, you might be able to assemble a series of 12 4-hour workshops given once a month, which, when added together, will become a complete farmers' soil and water conservation program. These workshops could be reinforced throughout the month with follow-up field visits.

Your first workshop could be a four-hour participatory mapping exercise so that your farming community can identify what the soil and water challenges they are facing are—and how serious they are. The map will also give them a baseline starting point from which they can improve. The second workshop could be on simple field tests to determine soil quality, and the third workshop could be on the importance of adding organic material to the soil in their fields.

Having made that decision, my next step would be to make technical and cultural decisions that would be appropriate to my farmers' context, and I would outline a workshop using the activities that I selected from the handbooks that I had reviewed (in this case, handbooks on participatory mapping). I would then write a three-quarter page summary of the step-by-step process of how to facilitate the workshop.

Drawing a How-to Card

Next, I would draw a series of very simple line drawings illustrating the techniques found in the summary and would print those up, back to back with the short field guide, into a single sheet handout. Drawings and illustrations should be appropriate for and familiar to the community members. One suggestion is to make simple sketches of what you would like to have on your how-to card—and then have one of the community members draw them up properly. This can help ensure that the illustrations are meaningful to the course participants. People in the illustrations will be wearing the appropriate clothing, and houses and fruits and vegetables will be contextual. If they are black-and-white drawings, they can be photocopied inexpensively.

Writing a Workshop Lesson Plan

I would then copy and paste from the three-quarter page summary into a lesson plan that would guide me through leading the workshop. The step-by-step lesson plan will have a specific schedule and detailed exercises and activities. The lesson plan will also focus on two-way communication.

Why write a lesson plan? The lesson plan will help keep you on track and on time if distractions happen in your workshop. You can always look back at your lesson plan to see where you were and what your next step should be. Plus, lesson plans represent for your organization the opportunity of replicability: if your project is successful, your organization may well want to replicate it in other communities. If you have a field guide and a lesson plan for a successful activity, they could be shared with your teammates so that they could quickly get to work as well.

Preparing for a Workshop

I would begin organizing a workshop itself two weeks or more in advance. Make sure that you have all your materials—such as large sheets of paper, and pens and markers for drawings. If the workshop lasts longer than two hours, you may also need to plan snacks, drinks, or a lunch.

Have two or three colleagues accompany you to help. This will be especially useful if you decide to break the participants down into sub-groups (for example, men and women, or teenagers and parents). If you are considering providing snacks, drinks or a lunch, put someone in charge so that you aren't distracted with the details and are free to focus completely on facilitating the workshop.

Review the lesson plan and workshop materials with your team in advance and make adaptations to the exercises so they are specific to your community context. You may choose to produce an illustrated poster for the workshop—especially if some participants cannot read. Role-play the exercises with your colleagues so that you will be better prepared when you present the workshop, and so you can discover if there are any cultural or linguistic problems.

Be Sure to Take Photographs

Put someone in charge of photos. Have them take the following shots:

- close-up detailed shots of participants;
- close-up detailed shots of the tools and materials that you use;
- photos of interesting drawings that you might have done on the newsprint;
- shots of the whole group;
- a few shots of you facilitating the workshop.

A big challenge that I see with non-profits organizations is a shortage of good quality photographs. Get close-ups of people so that they fill the frame of the photograph. Make sure they're smiling and looking at you. Make sure the lighting is good and that the photograph is in focus. If you're taking photographs outside where there is lots of glare, use your flash, it will balance out the glare. Take close-up pictures of people doing things—action shots. Take pictures of people doing the exercises in the workshop—such as drawing a map of the community—get right in there up close to take the photograph.

Don't take too many long distance shots because it is difficult to understand what they're about. Don't take pictures of the back of people looking at something—take pictures from the front, of people looking at something. When you're composing your photograph, make sure that there are not a lot of busy things happening in the background that will distract from your subject.

How to Facilitate a Participatory Workshop

A facilitator is not a teacher. During a participatory assessment, all participants are equally important. It is especially important to include women and marginalized members of the community. If you sense that, for example, men's comments may overshadow women's input, consider holding two meetings: one for men and one for women.

The facilitator should not lead the participants in a direction where he or she thinks they should go. This could be challenging for many facilitators who work with organizations that have a specialized mission, such as health, for example. What if the community feels that developing a market link for their agricultural projects is their highest priority—and that improved health isn't? In a purely participatory assessment a facilitator will need to set their mission aside and instead help the participants to better understand their own situation. They can then identify and prioritize challenges and participate in choosing and managing solutions and activities for improvement.

Facilitators need to maintain a positive attitude, and be open to new perspectives and new ideas. This may mean setting aside personal biases and assumptions. For example, in a participatory needs assessment workshop, your role as the facilitator is to help community members to do the following:

- identify needs, challenges, and problems faced by the community at large;
- prioritize these needs and challenges;
- discuss the needs and challenges with the highest priority to identify consensus and disagreement;
- discuss what the underlying causes of challenges may be;
- discuss potential solutions and identify local coping strategies they are currently using.

As the facilitator of a needs assessment you should *not* do the following:

- act as if you are the director of the group;
- give information: rather, let the participants hold discussions in order to provide you with information;
- let your own personal biases lead you to make assumptions about what is the correct outcome of a discussion.

Facilitators may need to draw upon new and unusual social skills, such as dealing with dominant personalities in group settings, while at the same time fostering the participation of silent participants.

Overview of Approaching the Members of a New Community for the First Time

If you are approaching a new community for the first time, it's best to approach community leaders initially, let them know the purpose of your visit, and help them to understand the importance of your work. It is a good idea to have introductory materials that describe your organization, the kind of work that you do, and illustrate the positive results that you have had. You might also want to explain about your method of engaging community members early on in the process—with the first step being a participatory needs assessment. Let them know that the purpose of the needs assessment is to get to know the people better, and to better understand their lives, needs and challenges. Be careful: these village leaders may feel that they can give you all of the information that you need. If that is the case, suggest two participatory meetings—one with them—and a second one with community members in the greatest need.

With the leaders' help and support, you will be able to meet community members interested in participating in your project—and in the initial needs assessment. Ask if they can help set up a 3-hour meeting with 10–12 community members. Suggest that you would like to meet with community members who represent the ultimate beneficiaries (mothers, fathers, families, farmers, weavers—whoever describes the community you want to work with). The size of the group should not exceed 12 people.

Communities are diverse and you need to be sure that you are working with a representative example of its members. Each sub-group of community members will have their own set of needs; some members may even be self-serving. You will need to choose which groups will be the most representational of overall community need.

FIELD GUIDE 10.1: PARTICIPATORY COMMUNITY NEEDS ASSESSMENTS

In community based adaptation, the adaptation activities you want to work with are grass-roots solutions for addressing community-identified need. In this first community based workshop you are going to use a Participatory Learning and Action (PLA) ranking tool. In the process workshop, participants voice different problems, challenges and needs they experience in the community—and then vote on them with voting tokens (small stones or beans) to prioritize them. You will use drawings to illustrate community-identified needs so that illiterate community members can participate equally in this process.

Workshop Preparation

Working with your community contacts, set up a 3-hour meeting with 10 or 12 community members two weeks or more in advance. Communities are diverse and you need to be sure that you are working with community members who represent the ultimate beneficiaries (mothers, fathers, families, farmers, weavers—whoever best describes the community you are intending to work with). Avoid basing your assessment on a meeting with people in higher positions: mayors or city council members, for example. It is important that women and marginalized members have a voice in the process.

Each sub-group will have their own set of needs; some members may even be self-serving. Ensure that everyone in the workshop is given an equal chance to voice the challenges that they see in the community. If there are cultural norms which may prevent some participants from speaking out, you may elect to form two groups out of the group of participants—for example, one of women separate from one of men—so that women can feel comfortable participating in the discussion. You could choose to hold a separate assessment with marginalized members of the community so they will feel comfortable speaking.

Review the lesson plan with your team and adapt the activities so they are specific to your community context. You may choose to produce an illustrated poster for the workshop—especially if some participants cannot read. Role-play the activities with your colleagues so that you are better prepared to present the workshop, and so you can discover if there are any cultural or linguistic problems. In the workshop you need to be simply a facilitator and not color the needs assessment with your own preferences.

Make sure that you have all the materials that you may need, such as notebook paper, large sheets of newsprint, tape, and markers for drawings. Since this is a three-hour workshop you may also need to plan snacks and drinks. Have two to three colleagues accompany you to help. This will be especially useful if you decide to break the participants down into sub-groups. If you are considering providing snacks, put someone in charge so that you are not distracted with the details and are free to focus completely on facilitating the workshop.

Community Need from the Community Perspective

After initial rapport building with the group, explain that the purpose of the activities is to understand and learn about their community from their perspective. Ask the group to imagine and discuss the problems and needs that are faced by the community as a whole.

As each need is identified by a community member, begin making simple illustrations that represent their challenges on notebook-sized sheets of paper (you can bring a selection of typical drawings to reduce time spent drawing). An example could be that if there is a housing shortage, draw a little house. After the group has come up with a complete set of needs/problems, arrange the different illustrations side-by-side into a rectangle on the ground or on a table.

Prioritizing Need Democratically

Have everyone leave the workshop area. Give each one of the participants voting tokens: 10 or 15 slips of paper, beans, or grains of corn. For privacy during voting, only one person should go into the workshop area at a time to vote. They should select the needs which they as an individual feel are the most important. It is their decision if they want to put all 10 tokens on one drawing or if they want to distribute them around several different challenges.

Results Matrix

When the participants have finished voting, count the total tokens on each drawing and write up a prioritized list ordered by the number of votes each problem received—with the need that received the most votes at the top. This would be a good time for the participants to take a break so that you can take a few minutes alone with the list and to draw a two-column matrix on a sheet of newsprint that everyone can see. In the left column write down the individual needs in their prioritized order (or draw little pictures again) and in the right column write the number of votes each one received.

This is a good time for the participants to have an open discussion about the results of the vote. Plus, if there are any unrelated needs competing for the highest position, it would be a good idea to let the participants do a second prioritization. For example, there might be two health-related challenges near the top and two micro-enterprise challenges near the top as well. You can ask the community which project they would like to start with first—a health project, or a micro-enterprise project—so that you can keep your project simple and not be faced with managing two dissimilar programs at once.

Selection of Most Important Needs for Project Development

It's very likely that the list will be a disorganized mixture of needs, challenges, underlying causes, and grievances. Work with the group to connect challenges to their underlying causes on the matrix so that they can see the relationship. If the matrix does not have any underlying causes, this would be a good time to ask the participants what they feel the causes of the top priority challenges are. It is likely that they might have more background information about a problem than you do—so this can be quite helpful. This is a good time to discuss potential solutions which they may see, and reveal which local coping strategies they are already using.

Community Feedback and Verification

Conclude the meeting by summarizing the two or three challenges that the community placed as their highest priorities and their relationships with underlying causes. Ask for feedback of your summary in order to get verification from the participants. Use your best facilitation skills to make sure that no one has any questions or further comments.

VOICING NEED

VOTING

TOTALS

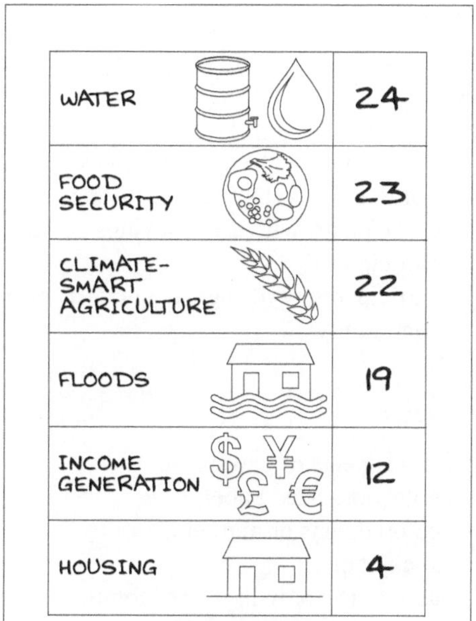

PRIORITIZE

Workshop Lesson Plan on Participatory Community Needs Assessment

- *Duration*: 3½ hours.
- *Background for lesson plan*: assumed knowledge. The community has agreed that a community needs assessment will be beneficial.

Anticipated Difficulties/Problems

- Information may not be well received due to cultural differences between workshop leaders and participants;
- community members may be resistant to behavioral changes;
- strong personalities may dominate workshops;
- stakeholders with vested interests may want a needs assessment that is favorable to them.

Solutions

- It is important that the facilitator is from a similar culture and speaks the same language as the participants;
- drawings and illustrations should be appropriate for and familiar to the community members;
- strong personalities may be able to be reasoned with in order to understand the importance of the participatory process. They should be given projects that will occupy them, or else taken aside for a VIP interview.

Purpose

The purpose of the workshop is for community members to assess and to prioritize needs within their community:

- participants will work in an environment where they feel safe discussing community needs and developing a needs list;
- participants will use drawings to identify needs so that non-readers will be able to participate;
- participants will agree upon a prioritized list of needs most important to the community.

Materials

- large sheets of newsprint;
- notebook-sized paper;
- colored pens or markers;
- sticky tape;
- about 15 drawings of community needs traditionally identified by community members;
- beans, slips of paper, or grains of corn for voting tokens.

Activity 1: Introduction. Duration 30 Minutes

The purpose of the activity is:

- to help build a feeling of team spirit and mutual understanding;
- to generate group self-esteem and creativity;
- to understand the purpose of the workshop.

Guided Practice

1 Introductions: ice breaker. Sing a song or play a game.
2 Tell the participants what they'll be able to do as a result of the lesson. Use the how-to card or a poster to visually show them the process:

   a discuss community need;
   b vote on the needs;
   c total the votes;
   d prioritize the needs;
   e discuss underlying causes and potential solutions.

Activity 2: Open Discussion about Community Need and Prioritizing Challenges and Needs. Duration 1½ Hours

The purpose of the activity is: participants will openly discuss problems and needs within the community.

Guided Practice

1 Ask the group to imagine and discuss the challenges and needs faced by the community.
2 As each need is identified by a community member, begin making simple illustrations that represent the challenges on notebook-sized sheets of paper (you can bring a selection of typical drawings to reduce time spent drawing).
3 After the group has come up with a good set of needs/problems, arrange the different illustrations side-by-side in a rectangle on the ground or on a table.
4 Have everyone leave the voting area and give each individual in the group 10 or 15 beans, slips of paper or grains of corn.
5 Ask them to place the voting tokens on the challenges they feel are the most important. It is their decision if they want to put all 10 tokens on one drawing or if they want to distribute them around several different challenges.
6 When the participants have finished voting, count the total tokens on each drawing and write up a prioritized list ordered by the number of votes each challenge received—with the need that received the most votes at the top. This would be a good time for the participants to take a break so that you can take a few minutes alone with the list and to draw a two-column matrix on a sheet of newsprint that everyone can see. In the left column, write down the individual needs in their prioritized order (or draw little pictures again) and in the right column write the number of votes each one received.

Have a break of 15 minutes.

Activity 3: Discovering Underlying Causes to Problems and Needs. Duration 1 Hour

The purpose of the activity is: participants will discover underlying causes to problems to challenges and needs.

Guided Practice

1 Review the results of the prioritized matrix with the group.
2 Ask them if the top items on the prioritized list represent the needs that are most important for the community.
3 If there are competing items at the top of the priority list, ask them to choose the ones that they would like to include in the first project.
4 Can they determine the underlying causes of the problems? List these underlying causes beneath the prioritized challenges that will be focused on during the first project. Frequently underlying causes are already on the priority list, simply move them into position beneath the top priorities.
5 If the matrix does not have any underlying causes, this would be a good time to ask the participants what they feel the causes of the top priority challenges are.
6 Discuss potential solutions which they may see and what local coping strategies they are already using.

Take 5 minutes to discuss the prioritized list, the problems chosen with the highest priority, and the underlying causes. Have participants talk about what they do and don't understand, what they do and don't like.

Activity 4: Conclusion. Duration 30 Minutes

Purpose

To reinforce what has been learned.

Guided Practice

1 discuss and review what has been learned and decided;
2 summarize the two or three challenges that the community places as their highest priority;
3 discuss the relationship between the challenges and their underlying causes;
4 review potential solutions to the community's challenges;
5 ask the group to ensure the validity of the summarized information;
6 make sure that there are no more questions and that everyone understands what happened in this process.

FIELD GUIDE 10.2: PARTICIPATORY CAPACITY AND VULNERABILITY ASSESSMENTS

Schedule a workshop with the community for six hours. You can also do this workshop in two three-hour sessions.

Activity 1: Seasonal Calendar. Duration 1½ Hours

This first activity in the workshop will be drawing a seasonal calendar in the form of a matrix. Draw a matrix on a sheet of newsprint—or several sheets taped together. On this calendar you're trying to establish relationships between times of the year, seasonal events, and special events that happen in the community.

- the rainy season;
- the dry season;
- periods of drought;
- extreme weather events;
- flooding;
- important livelihood activities;
- diseases;
- periods of hunger;
- planting and harvesting.

Along the top axis, you can write the initials for the 12 months of the year. So that all workshop participants can engage in the activity you can make it very visual by drawing little symbols—such as planting corn—so that non-readers will be included. Along the vertical axis on the left you can begin writing down the events that community members describe. Then, adjacent to the event you can make a mark in the appropriate months that the event occurs. One helpful technique is to have a second piece of newsprint that you can quickly write down ideas as the community comes up with them. This will give them the freedom to speak openly and quickly if they like. After a good number of ideas have been voiced, you can organize the ideas and the key events, since many of the individual ideas may be related or just phrased in a different manner. After having organized the events, you can then add them to the clean version of the calendar.

When everyone is satisfied that the calendar is accurate, introduce the following questions:

- Are the hazards concentrated in one time period?
- Are there time periods in the year which are the most difficult for community members and for their livelihoods?
- What are the community members' current coping strategies for dealing with these difficult periods?
- Capacity building: Which of the difficult periods are they having trouble coping with due to a lack of strategies?

Activity 2: Hazard Mapping. Duration 1½ Hours

The second 1½-hour activity in the workshop will be drawing a participatory hazard map of the community. Participatory mapping is a tremendous tool because all workshop participants can engage in the activity since it's very visual—non-readers will be included. Consider returning to the village the day before the workshop to tour the farm fields, forests, and water sources with one of the villagers. Take a few minutes to talk to farmers you meet in order to gain a greater understanding of the challenges they are facing.

Focus the exercise on drawing a community map on a sheet of newsprint—or several sheets taped together—in order to understand the spatial relationships between the different parts of the community. On this first map you're trying to establish relationships between major community components. How the village relates to the farm fields, hills, roads and where sources of water are.

When everybody at the workshop is satisfied that the basic map represents the community, farming areas and surrounding environmental resources, you can begin marking things on the map such as where individual homes are, or where their farm fields are. It's a good idea to represent buildings and farmers' plots using pieces of colored paper that can be attached to the map with removable tape so they can be moved or adjusted.

When everyone is satisfied that the map is accurate, introduce the idea of hazards that the community either suffers from or is at risk of suffering from in the future (Table 10.1). These hazards could include areas affected by extreme weather events, floods, heavy rainfall, drought, and health challenges. Once the hazard-prone areas have been indicated on the map, raise the following questions:

- Are the hazards concentrated in one area of the community?
- What negative impacts will the hazards have on community members and their assets?
- Who in the community is the most at risk from the hazards?
- Are there safe places in the neighborhood where community members can shelter from the hazards?

Note the following:

- What are the community members' current coping strategies for dealing with these difficult events?
- Capacity building: Which of the difficult events are they having trouble coping with due to a lack of strategies?

Table 10.1 Typical assets, resources, and hazards

Important assets and resources may include:	*Typical hazards may include:*
Income generation from agriculture	Extreme weather events such as hurricanes or cyclones
Crop land	Drought/heat waves
Livestock	Unpredictable beginning and end to the rainy season
Irrigation systems	Erratic rainfall or more or less rainfall
Health	Lack of water
Food reserves/food security	Shortage of food at specific times of the year
Environmental resources such as forests and water	Flooding
	Change in the timing of the growing season
	Health issues/disease

Activity 3: Historical Timeline. Duration 1½ Hours

The historical timeline is one that is a very simple matrix with years in the left column and important events in the right column. You are looking for insights into past hazards and events, and how they may have changed, intensified, or become more frequent over time.

These could include hurricanes, droughts, epidemics, famines, or floods. Other examples could include storms, erratic rainfall, a change in the timing of the growing seasons, and water availability. Hopefully, there will be village elders in the workshop that will allow you to get a long-term perspective from 20 or 25 years ago so that you and the villagers can see if these events are occurring more frequently.

Next, when the group has completed the timeline, introduce the subject of climate change into this timeline. Have they seen a change over time with changes in climate? When did they start noticing the changes? Some examples:

- Beginning 20 years ago, rainfall began diminishing; by how much?
- Beginning 20 years ago, the growing season changed; its shorter now—or it starts later.
- Beginning 20 years ago, the number of storms have increased and there is flooding when there didn't used to be.
- Beginning 20 years ago, they've had to walk progressively further to get water. How much further?

Briefly describe the changes they have seen, how they've changed and over what time frame. Does the community realize this is linked to climate change and realize that this may be ongoing and intensifying?

Note the following:

- What are the community members' current coping strategies for dealing with these difficult challenges?
- Capacity building: Which of the challenges are they having trouble coping with due to a lack of strategies?

Activity 4: Vulnerability Matrix: Impact of Hazards on Livelihood Assets and Resources. Duration 1½ Hours

The vulnerability matrix is another very simple matrix with important livelihood resources and assets in the left column and important hazards in the first three activities in the top row. It's a good idea to prepare a blank matrix on newsprint in advance. Doing a preliminary matrix on a sheet of notebook paper is also a good idea—you can then just organize and transfer the information onto the clean, blank matrix.

Once the matrix has been drawn with what the community feels are the most important hazards along the top row and assets and resources which are the most vulnerable along the left column ask them to rank in terms of importance which hazards are having the greatest impact on which resources. There are two ways that you can do this. Much like with needs assessment, you could give each participant 15 counting stones, lay the matrix on the floor and let them vote. Or, you can hold a discussion and let them rank the importance of hazard impact on resources and mark it on the matrix.

3 = greatest impact on the resource
2 = median impact on the resource
1 = low impact of the resource
0 = no impact of the resource

Note which hazards they are facing and prioritize them by which are the most challenging for them. Note which areas they feel the most vulnerable in and prioritize them.

At the end of this exercise you will have a matrix that prioritizes which hazards are causing the greatest risk and vulnerability for which livelihood assets and resources. As a concluding exercise, discuss this prioritization with the participants and make sure that they are in agreement.

EVENT	J	F	M	A	M	J	J	A	S	O	N	D
DRY SEASON	X	X	X	X							X	X
PLANTING				X	X					X		
RAINS					X	X	X	X	X	X		
DROUGHT								X	X			
HUNGER							X	X	X	X		
HARVEST			X								X	
FLOODS					X	X				X		
SCHOOL	X	X	X	X	X	X		X	X	X		

SEASONAL CALENDAR

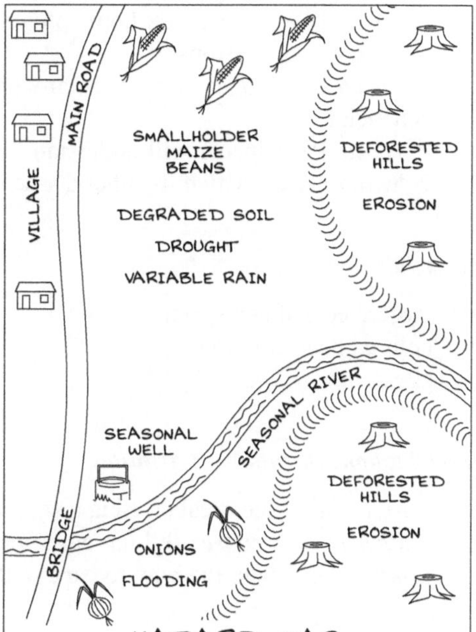

HAZARD MAP

YEAR	HAZARDS & EVENTS
2011	TROPICAL STORM – S' RAIN
2010	HURRICANE AGATHA
2009	DROUGHT AUG/SEPT
2008	SHIFT IN RAINY SEASON
2005	HURRICANE STAN
2004	FAMINE–POLOCHIC
2001	REFUGEES ARRIVE
2000	HURRICANE GORDON
1996	END OF CIVIL WAR
1995	HURRICANE MITCH
1976	7.6 EARTHQUAKE

HISTORICAL TIMELINE

RESOURCES ASSETS	SEASON CHANGE	DROUGHT	VARIABLE RAIN	TROPICAL STORMS
FOOD SECURITY	3	2	3	1
INCOME	3	3	3	2
AGRICULTURE	3	3	3	2
WATER ACCESS	0	3	3	1

HAZARD IMPACTS ON ASSETS

Workshop Lesson Plan for a Participatory Capacity and Vulnerability Assessment

- *Duration of workshop*: six hours plus a lunch break. You can also do this workshop in two three-hour sessions.
- *Assumed knowledge*: The community has agreed that a vulnerability assessment will be beneficial.
- *Purpose*: Participants will understand which hazards are causing the greatest risk and vulnerability for which livelihood assets.

Materials

- large sheets of newsprint;
- colored pens or markers;
- sticky tape.

Introduction. Duration 15 Minutes

- *Introductions*. Ice breaker: Sing a song or play a game.
- *Statement of purpose*: Tell the participants what they'll be able to do as a result of the lesson. Show them the how-to card so they can see the entire process.

Activity 1: Seasonal Calendar. Duration 1½ Hours

Purpose

On this calendar you will be establishing relationships between times of the year, seasonal events, and special events that happen in the community.

What to Do

Draw a matrix on a sheet of newsprint. Along the top axis, write the initials for the 12 months of the year. Along the vertical axis on the left you can begin writing down the events as community members identify them. Then, adjacent to the event, make a mark in the appropriate months that the event occurs.

Guided Practice

1 Ask the group to imagine and discuss important events that happen during the year in their community. These could include the agricultural season, the rainy season and the dry season, extreme weather events, periods of hunger, and periods of illness. It can also include community events such as the school year and annual festivals.
2 Once the calendar is complete, ask about periods of hazards such as flooding. Ask the following questions:

    a Are the hazards concentrated in one time period?
    b Are there periods in the year which are the most difficult for community members and their assets?

What are the community members' current coping strategies for dealing with these difficult periods? Capacity building: Which of the difficult periods are they having trouble coping with due to a lack of strategies?

Take a break for 15 minutes.

Activity 2: Hazard Mapping. Duration 1½ Hours

Purpose

Participants will discover spatial relationships between different components in the village and how hazards impact different areas and groups of people in the village.

What to Do

Consider returning to the village the day before the workshop to tour the farm fields, forests, and water sources with one of the villagers. Take a few minutes to talk to farmers you meet in order to gain a greater understanding of the scale of the community and to get a better sense of some of the challenges they are facing.

Tape several sheets of newsprint together and place them on the floor. A suggestion is to begin with black to draw the basic outline of the village, roads, pathways, and major farming areas. You can then use different colors for houses, rivers, and farm fields.

Guided Practice

1. Another suggestion is to take 15 minutes and have community members draw a small preliminary map of the village, surrounding farm fields, roads and pathways, farm fields and watersheds on a single sheet of newsprint. This will give you two things: you might discover that there is a good illustrator in the group—and you can quickly solve spatial problems by adjusting lines.
2. Let your group artist transfer this basic outline of the community onto a larger piece of taped-together sheets of newsprint.
3. Take colored sheets of paper and cut them out to represent additional features. These could be individual farm fields, they could be houses, schools, and churches—and stick them to the map with removable tape.
4. When everyone is satisfied that the map is accurate, introduce the idea of hazards that the community suffers from. These hazards could include extreme weather events, floods, heavy rainfall, drought, and health challenges. Once the hazards have been indicated on the map, introduce the following questions:

    a. Are the hazards concentrated in one area of the community?
    b. What negative impacts will the hazards have on community members and their assets?
    c. Who in the community is most at risk from the hazards?
    d. Are there safe places in the neighborhood where community members can shelter from the hazards?

What are the community members' current coping strategies for dealing with these difficult periods? Capacity building: Which of the difficult events are they having trouble coping with due to a lack of strategies?

Take a break for lunch: 30 minutes.

Activity 3: Historical Timeline. Duration 1½ Hours

Purpose

You are looking for insights into past hazards and events, and how they may have changed, intensified, or become more frequent over time.

What to Do

Draw a very simple matrix with a column for years on the left, and a column for important events on the right. Ask some of the elders of the group about a major event that occurred in the community—and what year it was. These could include hurricanes, droughts, epidemics, famines, or floods. Hopefully, there will be village elders in the workshop that will allow you to get a long-term perspective from 20 or 25 years ago so that you and the villagers can see if these events are occurring more frequently.

Guided Practice

1. Ask the group if they can remember major disasters that have happened over the past 25 to 30 years—and the year.
2. Begin organizing a matrix by year and event. A preliminary matrix can be drafted and then transferred.
3. When the group has completed the timeline, introduce the subject of climate change. Have they seen a change over time with of changes in climate? When did they start noticing the changes? Some examples:
    a. Beginning 20 years ago, rainfall began becoming less; by how much?
    b. Beginning 20 years ago, the growing season changed; it is shorter now—or it starts later.
    c. Beginning 20 years ago, storms have increased and there is flooding when there didn't used to be.
    d. Beginning 20 years ago, they've had to walk progressively further to get water. How much further?

Briefly describe the changes they have seen, how they've changed and over what timeframe. Does the community realize this is linked to climate change and realize that this may be ongoing and intensifying?

What are the community members' current coping strategies for dealing with these difficult events? Capacity building: Which of the difficult events are they having trouble coping with due to a lack of strategies?

Take a break for 15 minutes.

Activity 4: Matrix: Impact of Hazards on Livelihood Assets and Resources. Duration 1½ Hours

Purpose

To determine which community assets and resources are the most vulnerable and to determine which hazards are having the greatest impact on the community resources.

What to Do

Draw a simple matrix with livelihood resources and assets in the left column and hazards in the top row.

Guided Practice

1 Ask the group to name the most important hazards as identified in the past three exercises. Place those in the top row.
2 Ask the group to name their most important livelihood assets, and add those to the column on the left.
3 Rank which hazards are having the greatest impact on which assets. Much in the same way as the needs assessment, each participant could vote with 15 tokens. Or, you can hold a discussion and let them rank the importance of hazard impact on resources within the matrix with a "3" indicating the greatest impact on the resource and a "0" the least.
4 Note which hazards are creating the greatest challenges for them.
5 Note which are the areas where they feel the most vulnerable and prioritize them.

Activity 5: Conclusion. Duration 1½ Hours

Purpose

To reinforce what has been learned.

What to Do

Have the participants take 10 minutes to discuss the prioritized list. Discuss and review what has been learned and decided. Have participants talk about what they do and don't understand, what they do and don't like.

1 Ask the group to ensure the validity of the summarized information.
2 Make sure that there are no more questions and that everyone has understood the process.

FIELD GUIDE 10.3: PREPARING FAMILY GARDEN BEDS AND PLANTING SEEDS

How to Teach Community Members to Design a Simple Garden, Dig Beds, and Plant Seeds in a Workshop

Schedule a workshop with the community for eight hours. You can also do this workshop in two four-hour sessions.

Introduction to Nutrition and Family Gardens

For many people living in the cycle of poverty, the idea of starting a kitchen garden might seem overwhelming. It could be the time investment, it might be perceived costs. It might be a lack of know-how: what to plant, how to plant, and how to care for a garden. However, the positive benefits make it worthwhile enabling community members to garden for nutrition purposes.

Start small, think simple. The purpose of the first year's garden is to give the participants a win—so that they will be encouraged to plant again the following year. Even if they plant only one bed, 1 meter by 4 meters, they should be able to get positive, delicious, nutritious results.

Planning the Garden

A garden must first be planned and designed. In the first year don't get into too much detail; don't scare people away from the idea with too much information. During the course of the year you can gradually teach them more so that they can do a better job of planning for year two.

Discuss planning for sun, exposure to wind, runoff, family size and food production, and crop choices for nutrition. Provide large sheets of paper for them to design an example garden. Ask participants to sketch the area around their house and begin thinking of a good location for their garden. Work with the family to make a decision: a single small bed the first year—or something bigger?

Discuss the Importance of Organic Material (OM)

Discuss the importance of organic matter for the soil and the beds. In the first year, since they may not have compost, let them know that they can begin by spreading whatever chopped-up organic material (OM) they can find on top of the staked-out bed location. This can be leaves, manure, corn stalks, vegetable-based kitchen scraps. Organic material in garden soil provides nutrients, structure and facilitates holding water. Explain how many freely available types of OM can be found around the village to get a garden plot started. Have participants discuss other materials that they might be able to use.

Raised Garden Beds

These provide a soft environment for roots, they drain well, and the soil flora and fauna receive the oxygen they need. Soil is a living, breathing organism of sand, clay, organic matter, earthworms, micro-organisms, nutrients, minerals, water, and plant roots. It can suffer from being too wet, too dry, and too sandy. Organic material and the soft soil of the raised beds are best for root penetration, drainage, aeration, nutrient availability, and structure.

Take Turns at Laying out and Digging a Bed, Mixing in Organic Material, and Smoothing the Bed

Stake out an area for a bed that is no wider than 1 meter; clear it of any vegetation or trash. Lay some organic material on top of the staked-out area. Beginning at one end of the new bed, dig a 1-meter-long trench one shovel in depth and one shovel in width wide. Place the soil to the side. With a garden fork or with the shovel, loosen the soil in the bottom of the trench a further shovel depth—but don't remove it. Place some more organic material in the trench.

Dig a second trench alongside the first one, tossing the soil into the first trench. With the addition of the OM and the fluffing of the soil, the soil should now be higher than the surrounding terrain. Continue this process for the entire length of the bed. Place the reserved soil from the first trench into the last trench.

Using a garden rake, and without walking on the new bed, carefully break up any clumps of soil and rake the surface of the new bed smooth, flat and level. Carefully rake the outer edges so that they slope at a 45-degree angle, and so that a small lip forms at the upper edge for holding water.

Lay Out a Seed Grid and Plant a Few Example Seeds at the Right Distances and Depths

Nutritious plants should have already been selected and seeds obtained. The seeds should come with directions for correct spacing for planting and correct planting depths. Explain practical examples for seed spacing and planting depth: Crowding of plants, wasted space, and seed size versus planting depth. Explain how to measure out and mark the beds for planting the seeds. Being sure not to walk on the new beds. Help them plant a few seeds by demonstrating making a groove or a hole in the soil, planting and then covering the seed with soil. Show how to label the seed rows. Let the participants lay out the beds for the rest of the seeds.

Gently Water the Newly Planted Seedbeds and Build a Fence

Use a watering can with a fine spray and gently water the newly planted seeds. Water slowly enough that the water can soak in and not form pools; pools can cause the seeds to float to the surface. Explain the best times of day to water, the frequency, the duration and the quantity. Let the workshop participants take turns watering the newly planted seedbeds.

Put together a simple fence to protect the bed from animals. Be creative and use any free materials just to get going the first year. The fence could be made of branches, old tires, old barrels, or old pallets.

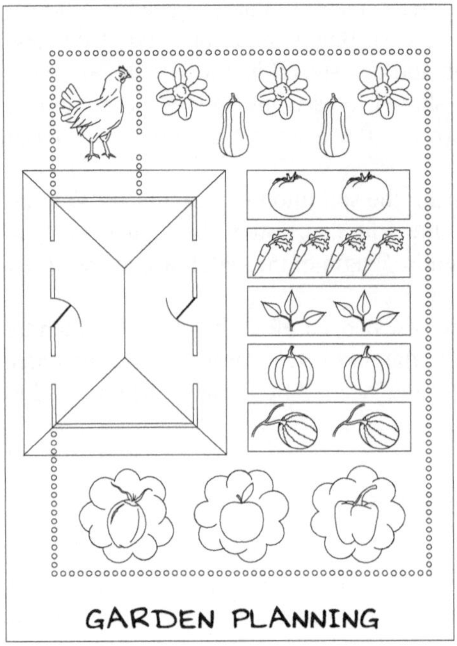

GARDEN PLANNING

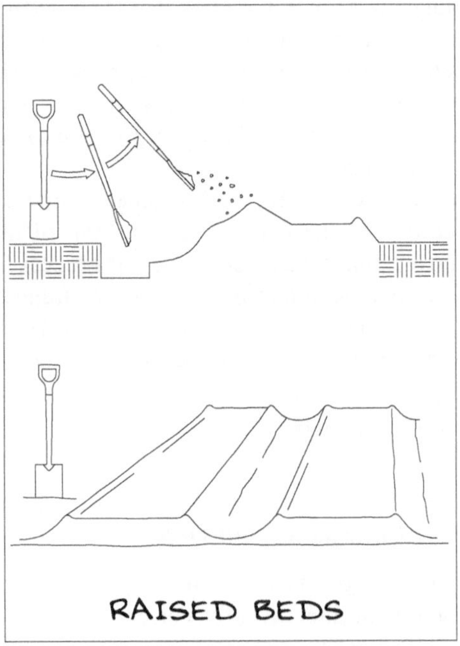

RAISED BEDS

PLANTING SEEDS/STARTS

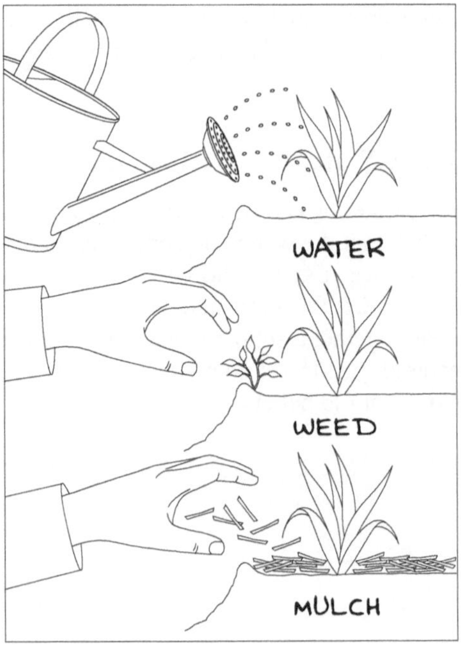

WATER
WEED
MULCH

Workshop Lesson Plan for Preparing Family Garden Beds and Planting Seeds

- *Duration*: eight hours including lunch (can be done in two half-day workshops).
- *Purpose*: support participants in planting their first garden to improve food security and family nutrition:
    - Objective 1: All will know the proper method for preparing a raised bed that includes organic matter.
    - Objective 2: All participants will successfully be able to properly plant a variety of seeds.
    - Objective 3: All participants will understand why clearing and fencing the garden plot are important.

Materials

- artist's drawings/posters; the scenes and people they contain should appear familiar to workshop participants;
- variety of seed to be provided for participants by the project;
- basic garden tools for instructor to use: machete, shovel, rake, trowel, watering can;
- a water source;
- organic material (leaves, crop residue, manure) for mixing with soil;
- how-to cards without written words for workshop participants to take home;
- large sheets of newsprint and tape;
- colored markers.

Preparation

- Find a workshop location. It can be the garden of a participant, the community center, the school or the church;
- plan out the demonstration garden; this will be good preparation and practice for when you present this workshop;
- clear the area of brush, weeds, and trash in advance of the workshop;
- prepare several beds in advance so there will be enough beds by the end of the day for planting seeds;
- work with an agriculturist/nutritionist as necessary on background information and selection of nutritious plants.

Activity 1: Garden Bed Introduction. Duration 90 Minutes (Including a 15-Minute Ice Breaker)

Purpose

Introduce and explain what the value of a properly planned and prepared garden is in relation to family health.

What to Do

1 Ice breaker: Introductions. Sing a song or play a game.

2 Introduction to workshop: Tell the participants what they'll be able to do as a result of the lesson.
3 Use practical examples of why they need to plan their garden:

   a the need to plan for sun, exposure to wind, runoff, family size and food production, and crop choices for nutrition;
   b use large sheets of paper to design an example garden;
   c ask participants to sketch their yard and begin thinking of a good location for their garden.

4 Summarize briefly:

   a examples of why the plot is cleared and fenced: residual contamination, weeds, insects, damage from animals;
   b examples of why they should loosen the soil and mix in organic material: Root penetration, drainage, aeration, nutrient availability, structure, micro ecosystem;
   c examples of seed spacing and planting depth: Crowding of plants, wasted space and seed size versus depth;
   d simple explanation of watering: Best times of day, frequency, duration and quantity.

Workshop Participants

Take 5 minutes to talk about what you do and don't understand, what you do and don't like.

Activity 2: Planning and Clearing a Small Area of the Plot and Looking at a Sample of Fencing. Duration 30 Minutes

This part of the lesson needs to be held outdoors in the example garden.

Purpose

Show how the best spot was chosen for this garden and discuss fencing and how the land was cleared.

What to Do

1 Discuss the location of the beds in relation to space, exposure, sun. What decisions were made?
2 Ask for feedback from the participants.
3 Show what was cleared off of the garden plot and re-emphasize why.
4 Show a sample of the fencing that was used.

Workshop Participants

Take 5 minutes to talk about what you do and don't understand, what you do and don't like.
Take a break for 15 minutes.

Activity 3: Looking at the Organic Material Samples Collected from Around the Village. Duration 15 Minutes

Purpose

Explain how many freely available types of OM are available around the village for getting a garden plot started.

What to Do

1 discuss the different materials and where they were found;
2 have participants discuss other materials that they might be able to use.

Activity 4: Taking Turns Laying Out and Digging a Bed, Mixing in Organic Material and Smoothing the Bed. Duration 90 Minutes

Purpose

To practice the digging technique and the shaping of a raised bed.

What to Do

1 show how to measure and stake out a bed—then let participants lay out two beds;
2 show how to double dig the bed and add in the OM—then let the participants take turns digging the bed;
3 show how the final shape of the bed can be formed and smoothed with a rake.

Take a break for lunch: 30 minutes.

Activity 5: Laying Out a Seed Grid and Planting a Few Example Seeds at the Right Distances and Depths. Duration 60 Minutes

Purpose

To re-emphasize seed spacing and planting depth.

What to Do

1 show a chart with spacings for the different seeds and demonstrate how to transfer that to the planting bed;
2 let the participants finish laying out the beds for the different seeds;
3 the chart gives seed planting depth; demonstrate making a groove or a hole in the soil, planting and then covering the seed;
4 show how to label the seed rows.

Independent Practice

Activity 6: Planting the Rest of the Seeds and Labeling the Rows. Duration 90 Minutes

Purpose

To let participants work through the entire process themselves—and to reinforce what has been learned.

What to Do

1 let participants finish planting the seeds in the beds;
2 let them label the seed rows.

Workshop Participants

Take 5 minutes to talk about what you do and don't understand, what you do and don't like.

Activity 7: Gently Water the Newly Planted Seedbeds. Duration 15 Minutes

Purpose

To let participants work through the entire process themselves—and to reinforce what has been learned.

What to Do

1 show the proper technique for watering the new seedbeds;
2 let the workshop participants take turns watering the newly planted seedbeds.

Workshop Participants

Take 5 minutes to talk about what you do and don't understand, what you do and don't like.

Activity 8: Conclusion: Principles of Plot and Bed Preparation, Organic Material, Seed Planting and Watering. Duration 30 Minutes

Purpose

To reinforce what has been learned and to discuss common mistakes and positive solutions.

What to Do

1 discuss and review what has been learned;
2 discuss common mistakes and positive solutions observed during independent practice;
3 reinforce the principles of garden planning and layout:

a plan the garden for location, size, exposure and nutritional crops;
b lay out, double dig and add OM to the beds;
c form with a rake;
d plant seeds to correct spacing and depth;
e water.

Workshop Participants

Take 10 minutes to talk about what you do and don't understand, what you do and don't like.

Homework

Participants are to prepare a garden plan for their individual gardens for the first home visit by the field staff.

FIELD GUIDE 10.4: SOIL RESTORATION AND CONSERVATION FOR SMALLHOLDER FARMERS

Worldwide, challenges for smallholder farmers have increased. Harvest production may be reduced, leading to lower incomes and fewer crops for family consumption. These challenges can be due to depleted soils, lack of funds to purchase fertilizer, changes in the beginning and end of the rainy season, unpredictable rain during the rainy season, and increased soil erosion and crop damage during extreme weather events. There are simple, low-cost/no-cost activities that subsistence farmers can adopt that can increase harvest production by restoring soil, reducing the need for chemical fertilizers, buffering the effects of variable rainfall, and protecting valuable topsoil from erosion, thereby increasing family nutrition and agricultural income.

What is Soil?

Soil is a living, breathing organism of clay, organic material, earthworms, microorganisms, beneficial flora and fauna, nutrients, minerals, water, and plant roots. Plant growth can suffer if soil is too wet, too dry and too sandy, too clayey, too exposed, and too steep.

Soil Moisture

Moisture in the soil is a chief determinant in crop growth and in agricultural production. Soil moisture improves soil chemical processes and also acts as a transport mechanism for getting nutrients to the plants. Moisture stored in the soil provides a buffer during dry periods or during periods of unpredictable rain. Sufficient organic material in the soil and mulch on the surface of the soil can help rainwater percolate into the soil in order to build up soil moisture.

Organic Material

Organic material decomposes in the soil and releases vital nutrients for crops—reducing the need to purchase expensive fertilizer. Increased organic material in the soil also helps to retain soil moisture for longer periods of time—buffering against unpredictable rain or an early end to the rainy season. Organic material in the soil is a benefit for root penetration, drainage, aeration, nutrient availability, soil structure, and can neutralize pH imbalances.

Discussing the Importance of Organic Material

In the first year, farmers may not have organic material. Let them know that they can begin by spreading whatever chopped-up organic material (OM) they can find on top of their field. This can be leaves, manure, chopped-up corn stalks, vegetable-based kitchen scraps. Explain how many freely available types of OM are available around their village; have participants discuss other materials that they might be able to use. Discuss the importance of re-incorporating maize stalks and other crop residues back into the soil rather than burning them prior to planting.

Farmers can spread organic material on top of their field as they prepare their fields prior to tilling. The organic material will mix in with their soil during tilling.

Mulching for Increased Organic Material, and Reduced Erosion and Evaporation

The addition of mulch to the top of the soil can reduce soil temperature, keep weeds down, improve drainage, attract earthworms, and reduce both wind and water erosion. It can be an excellent method of adding organic material to the soil as the mulch decomposes during the course of the growing season. It is excellent for water conservation: it reduces evaporation, protecting garden plants by retaining soil moisture when water is scarce. Adding mulch to your field is very simple. Use the same materials that you used for making compost: leaves, dry grass, rice stalks, straw, and other agricultural residues. Simply place a thin layer on the soil after planting seeds. As the plants begin to grow, add another layer until you have 5 to 10 cm.

Making Compost

Compost is the earthy, dark crumbly material that results from the decomposition of plant residue. It is rich in nutrients and organic matter and can be used as a plant fertilizer. To make compost, you need the right mix of organisms, air, water, and plant wastes such as grass clippings, food scraps, manures, leaf litter, and straw.

Compost can be made in a bin or simply as a pile approximately a meter square and meter high. Find a location for the compost pile that is well-drained and sunny. Unless you're lucky enough to have the materials that you need to make an instant compost pile, begin adding materials as you collect them to the top of the pile. It's good to alternate layers of dry things like leaves and straw with layers of green grass clippings and kitchen waste. A compost pile should be turned every two or three weeks to allow more air into the compost pile. Show the participants examples of finished compost so they know what it should look like.

Compost can be added to the field's surface before preparation for planting—in this way it will mix in with the field's soil during tilling and be accessible to the plants' roots. It can also be added to the surface of the field after planting and before the application of mulch. Its nutrients can then percolate into the soil with rainwater.

Conclusion

Even highly depleted soils can over the course of several years be restored to a vital condition. The addition of organic material and compost will increase the soil's ability to retain moisture, increase nutrients stored in the soil, increase beneficial microbes and flora and fauna, and will improve the structure of heavily compacted soil. Mulching will conserve restored soil by reducing moisture loss through evaporation, will contribute organic material and nutrients to the soil, and will prevent the loss of valuable topsoil by protecting the soil from wind and water erosion.

°C
EVAPORATION
EROSION

SOURCES OF ORGANIC MATERIAL

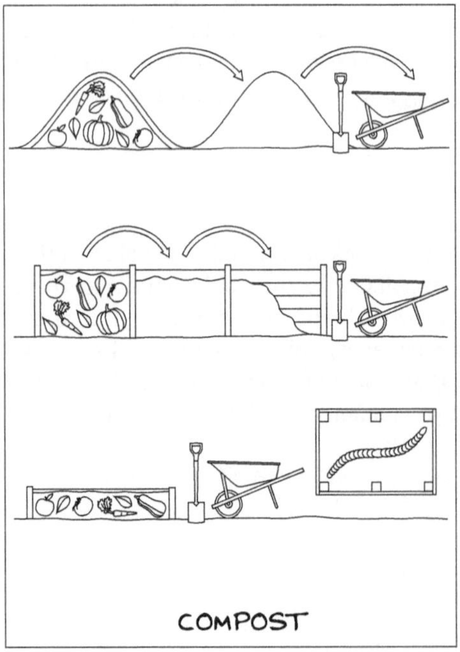
COMPOST

MULCH LIGHTLY AFTER PLANTING
MULCH AS PLANTS GROW

Workshop Lesson Plan for Soil Restoration and Conservation for Smallholder Farmers

- *Duration*: eight hours including lunch (can be done in two, half-day workshops).
- *Purpose*: support in learning about the importance of soil restoration and conservation:
    - Objective 1: All participants will understand what soil is composed of and the importance of moisture in soil.
    - Objective 2: All will know the importance of organic matter (OM) in soil and that that OM helps soil retain moisture.
    - Objective 3: All will understand that mulch can protect and conserve soil, and reduce evaporative losses.
    - Objective 4: Participants will understand how to build a compost pile.
    - Objective 5: All will understand how to apply mulches and incorporate organic material and compost into soil.

Materials

- artist's drawings/posters; the scenes and people they contain should appear familiar to workshop participants;
- basic garden tools for instructor to use: machete, shovel, rake, trowel, watering can;
- a water source;
- organic material (leaves, crop residue, manure) for mixing with soil, for mulching soil and for making compost;
- medium-sized pane of clear glass or plastic (~0.5 sqm);
- how-to cards without written words for workshop participants to take home;
- large sheets of newsprint and tape;
- colored markers.

Preparation

- Workshop location. A participant's farm field or a school yard cleared of brush, weeds and trash in advance.

Activity 1: Soil Restoration and Conservation Introduction. Duration 60 Minutes (Including a 15-Minute Ice Breaker)

Purpose

To understand the importance of healthy soil and the subsistence farm.

What to Do

1 Ice breaker: Introductions. Sing a song or play a game.
2 Introduction to workshop: Tell the participants what they'll be able to do as a result of the lesson.
3 Discuss the challenges that farmers face due to a lack of soil health on their farm fields.

4 Use practical examples of why farmers need to restore and conserve their farm's soil:

   a so that their crops can have the moisture and nutrients that they need for productive harvests;
   b so that the soil can be protected from wind and water erosion.

5 Discuss that there are low-cost/no-cost activities to improve challenges with soil.

Workshop Participants

Take 5 minutes to talk about what you do and don't understand, what you do and don't like.

Activity 2: What Is Soil? Duration 30 Minutes

Purpose

To understand what soil is composed of.

What to Do

1 discuss that soil is composed of sand and clay, organic matter, water, nutrients, and beneficial flora and fauna;
2 discuss how plants can suffer from soil being too wet, too dry, too sandy or clayey, too exposed, and too steep;
3 show samples of healthy soil and depleted soils.

Workshop Participants

Take 5 minutes to talk about what you do and don't understand, what you do and don't like.

Activity 3: Organic Material. Duration 30 Minutes

Purpose

To understand the importance of organic matter in soil including for retaining moisture.

What to Do

1 OM decomposes in the soil, releases valuable nutrients for crops and reduces the need to purchase fertilizer.
2 Discuss how organic matter helps soil retain moisture for longer periods of time—buffering against variable rain.
3 Discuss how organic material facilitates water penetration and drainage.
4 Discuss how organic material improves nutrient availability, soil structure, and can neutralize pH imbalances.

Workshop Participants

Take 5 minutes to talk about what you do and don't understand, what you do and don't like.

Take a break for 15 minutes.

Activity 4: Looking at the Organic Material Samples Collected from Around the Village. Duration 15 Minutes

This part of the lesson needs to be held outdoors in the example garden.

Purpose

To understand how many freely available types of OM are available around the village for getting started.

What to Do

1 show them the organic material samples collected from around the village;
2 discuss the different materials and where they were found;
3 have participants discuss other materials that they might be able to use.

Activity 5: Mulching for Increased Organic Material, and Reduced Erosion and Evaporation. Duration 60 Minutes

Purpose

To understand how mulch can protect and conserve soil and reduce evaporative losses.

What to Do

1 Demonstrate how mulching can reduce evaporation:

   a till and lightly moisten a small area of soil exposed to direct sunlight (before workshop);
   b mark off four small plots no more than half a meter square each (before workshop);
   c cover the first plot with a pane of glass raised about 10 cm above the soil surface. Seal the edges;
   d carefully and completely cover the second plot with chopped vegetation (straw, leaves, grass);
   e loosely cover the third plot with chopped vegetation;
   f leave the fourth plot uncovered and return to discussion.

2 Discuss how the addition of mulch to the topsoil can reduce the soil temperature, keep weeds down, improve drainage, add organic material, conserve water and reduce both wind and water erosion.
3 Return to the test plots after one hour. Moisture should have begun to collect on the underside of the glass pane. Have the participants discuss where the moisture came from. Discuss soil evaporation and its effect on soils—and plants. Remove the glass and have the participants feel the soil beneath—it should still be moist. Then have participants remove the mulch from test plots two and three. The soil should still be moist and plot two but less moist in plot three. Plot four, which was left uncovered should be even less moist and perhaps completely dry.

Take a break for lunch: 30 minutes—or end of the first half of workshop. Start the second half another day.

Activity 6: Making Compost. Duration 90 Minutes

Purpose

To understand how to build a compost pile.

What to Do

1 Compost is made from the decomposition of plant residue, is rich in nutrients, and can be used as a fertilizer.
2 Let the participants build a meter-square compost pile using the organic materials that you brought to the workshop.
3 Alternately layer dry things like leaves and straw with layers of green grass clippings and kitchen waste.
4 Gently water the compost pile so that it's moist.
5 Discuss how the compost pile should be turned every two weeks to allow air into its center.
6 Show participants an example of finished compost so they know what it should look like.

Workshop Participants

Take 5 minutes to talk about what you do and don't understand, what you do and don't like.

Activity 7: Incorporating Organic Material into Your Soil during Soil Preparation. Duration 90 Minutes

Purpose

To practice applying mulches and incorporating organic material and compost into soil.

What to Do

1 Show how farmers can spread organic material on top of their field as they prepare their fields prior to tilling. After planting, farmers can put another layer of finely chopped material on top of the freshly planted field.
2 Show how compost can also be added to the field's surface before preparation for planting—in this way, it will mix in with the field's soil during tilling and be accessible to the plants' roots. It can also be added to the surface of the field after planting and before the application of mulch. Its nutrients can then percolate into the soil with rainwater.
3 Show how adding mulch is very simple. Use the same materials that you used for making compost: leaves, dry grass, rice stalks, straw, and other agricultural residues and place a thin layer on the soil after planting seeds. Discuss how as the plants begin to grow, another layer can be added until you have 5–10 cm.

Take a break for 15 minutes.

Activity 8: Conclusion: Soil Restoration and Conservation. Duration 30 Minutes

Purpose

To reinforce what has been learned and to discuss positive solutions.

What to Do

1 Discuss and review what has been learned.
2 Reinforce the principles of soil restoration and conservation:
   a depleted soils can be restored with the addition of organic material and compost;
   b restored soils have increased nutrients and flora and fauna—and a greater ability to retain moisture;
   c restored soils can be conserved by applying mulches;
   d mulches reduce evaporation and contribute organic material and nutrients to the soil;
   e mulches reduce wind and water erosion.

Workshop Participants

Take 10 minutes to talk about what you do and don't understand, what you do and don't like.

FIELD GUIDE 10.5: PARTICIPATORY MAPPING OF SOIL AND WATER RESOURCES

Participatory mapping is an excellent way of learning in greater detail about the community, their resources, the hazards they face, and how the village, farm fields, roads, hills and water sources interrelate. It is also a method for the community members to see things they take for granted every day through a new lens. Participatory mapping is a tremendous tool because all workshop participants can engage in the activity—it's very visual—non-readers will be included.

Organize a 5-hour workshop with 12–15 farmers from your community. The purpose of this workshop is for the farmers to understand the impact that wind, soil health, location of water sources, variable rains, runoff, floods, and drought have on agricultural productivity.

Geography of the Community

Consider returning to the village the day before the workshop to tour the farm fields, forests, and water sources with one of the farmers. Take a few minutes to talk to farmers you meet in order to gain a greater understanding of the challenges they are facing. Work in advance with an agriculturist on background information on soil restoration, water conservation and management, and crop selection so that you are better prepared. Arrange for him to participate in a second, follow-up workshop for site visits, to discuss the results of the mapping exercise and to suggest solutions to the farmers' challenges.

Drawing the Basic Map

Tape several sheets of newsprint together and place them on the floor. A suggestion is to begin with a black marker to draw the basic outline of the village, roads, pathways, and major farming areas. You can then use different colors for houses, rivers, and farm fields. Another suggestion is to take 15 minutes and have community members draw a small preliminary map of the village, surrounding farm fields, roads, pathways, and watersheds on a single sheet of newsprint. This will accomplish two things: you can quickly solve spatial problems by moving things around on the preliminary drawing—and you might discover that there is a good illustrator in the group. Let your group artist transfer this basic outline of the community onto the larger piece of taped together sheets of newsprint.

Indicating the Location of Community Features

Take colored sheets of paper and cut them out to represent additional features. These could be individual farm fields, houses, and school. Stick them to the map with removable tape so they can be moved or adjusted; by removing these bits of paper completely, the map can be used again for a different assessment.

Features Important for Soil, Water and Agriculture

The following are important features:

- sources of both domestic and agricultural water and their relationship to the village/ farmlands rivers and streams;

- seasonal availability of both domestic and agricultural water;
- the location of steep hillsides or canyons;
- community land, forest boundaries, grazing/pasture lands.

When everybody at the workshop is satisfied, begin applying farm specific information. Examples could be:

- farmlands vulnerable to drought (or insufficient access to water);
- farmlands vulnerable to flooding, too much wind exposure, and other weather-related hazards;
- areas that suffer from excessive runoff;
- areas of high erosion and gullies;
- the location of different crops;
- farmers' perception of the fertility of their soil: good, medium, or poor;
- types of soil;
- areas of high and low agricultural productivity.

Indicating Community Hazards on the Map

Next is to begin the process of overlaying hazards impacting the community onto the map. These hazards might be floods, portions of the community that are most affected by drought, by heavy rain, or by extreme weather events. Which parts of the community, which people, which personal assets, which environmental resources, and which livelihoods are the most vulnerable to the hazards as identified on the hazard map?

Conclusion

To reinforce what has been learned

1. to discuss and review what has been learned;
2. to reinforce the challenges for farmers that were identified during the mapping exercise;
3. to discuss whether hazards such as floods, variable rainfall and drought impact soil and water resources;
4. to discuss whether the intensity of the hazards is increased by farming practices and deforestation.

What are the community members' current coping strategies for dealing with these difficult periods? Capacity building: Which of the difficult events are they having trouble coping with due to a lack of strategies?

Summarize a list of the challenges and hazards farmers face in preparation for the visit by the extension agent.

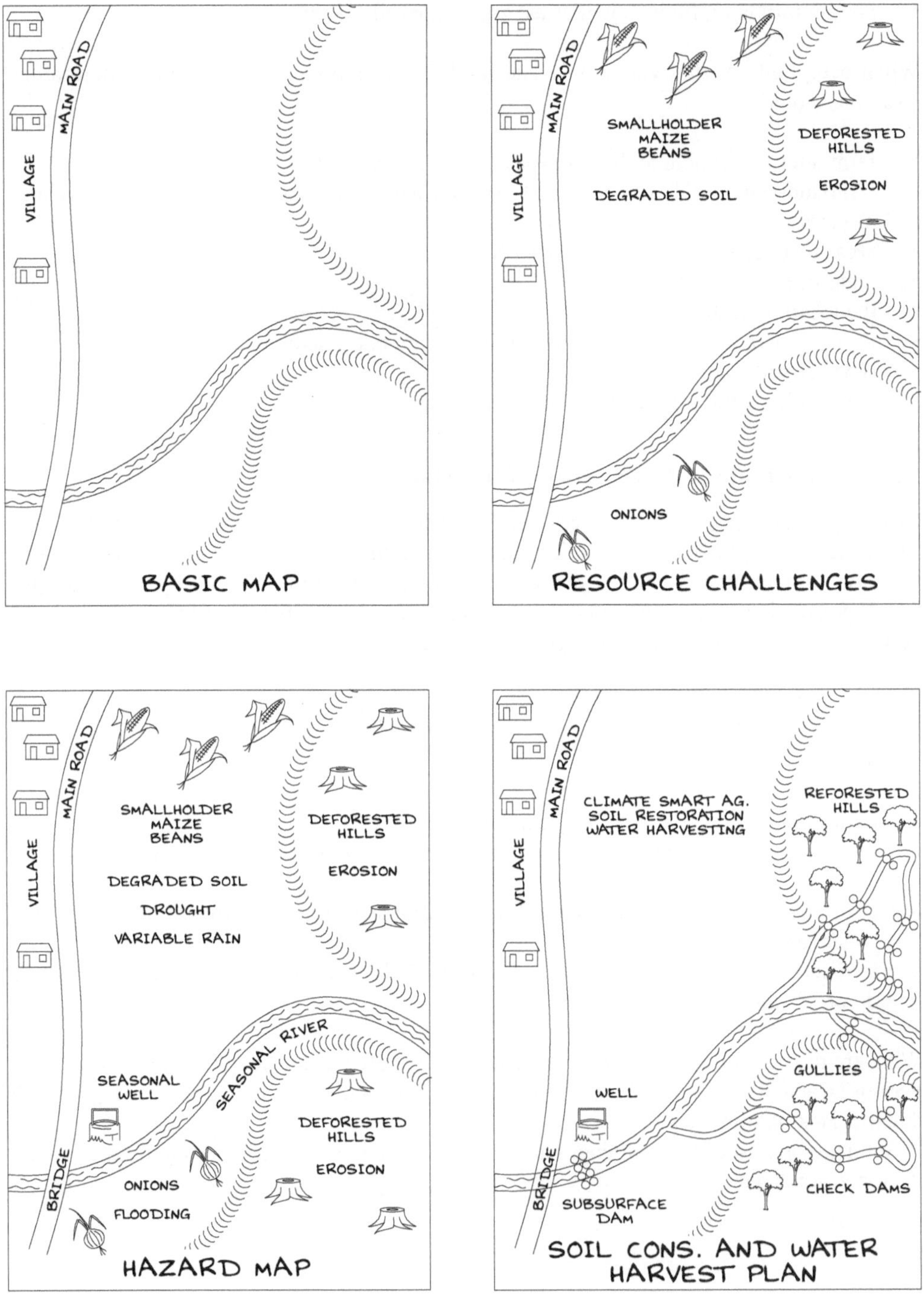
VILLAGE
MAIN ROAD
BASIC MAP
VILLAGE
MAIN ROAD
SMALLHOLDER
MAIZE
BEANS
DEGRADED SOIL
DEFORESTED
HILLS
EROSION
ONIONS
RESOURCE CHALLENGES
VILLAGE
MAIN ROAD
SMALLHOLDER
MAIZE
BEANS
DEGRADED SOIL
DROUGHT
VARIABLE RAIN
DEFORESTED
HILLS
EROSION
SEASONAL RIVER
SEASONAL
WELL
DEFORESTED
HILLS
EROSION
BRIDGE
ONIONS
FLOODING
HAZARD MAP
VILLAGE
MAIN ROAD
CLIMATE SMART AG.
SOIL RESTORATION
WATER HARVESTING
REFORESTED
HILLS
GULLIES
WELL
BRIDGE
SUBSURFACE
DAM
CHECK DAMS
SOIL CONS. AND WATER
HARVEST PLAN

Workshop Lesson Plan for Participatory Mapping for Soil and Water Resources

- *Duration*: five hours (this can be done in one two-hour workshop and a second three-hour workshop).
- *Purpose*: support participants in drawing a map of their farm fields, water sources, and locations of hazards so that the farmers can understand climate, water, and soil challenges to crop productivity:
  - Objective 1: All participants will understand why they are drawing a map.
  - Objective 2: All will know where domestic and agricultural and water come from.
  - Objective 3: All will understand climate, water, and soil challenges to crop productivity.
  - Objective 4: All will know the impacts that the hazards have on the community and on crop production.
  - Objective 5: All will know the origin of floodwater.

Materials

- artist's drawings/posters; make sure that the scenes and people they contain will appear familiar to participants;
- how-to cards without written words for workshop participants to take home;
- large sheets of newsprint and tape;
- colored markers;
- colored construction paper.

Preparation

- Find a workshop location: a community center, schoolroom, or church.
- Consider returning to the village the day before the workshop to tour the farm fields, forests, and water sources with one of the farmers. Take a few minutes to talk to farmers you meet in order to gain a greater understanding of the challenges they are facing.
- Work in advance with an agriculturist on background information on soil restoration, water conservation and management, and crop selection so that you are better prepared. Arrange for him to participate in a second workshop for site visits, to discuss the results of the mapping exercise and to suggest solutions to the farmer's challenges.

Activity 1: Map Introduction. Duration 30 Minutes

Purpose

Introduce and explain the value of drawing a map of farm fields, terrain, water sources, and crops.

What to Do

1 Introductions. Ice breaker: Sing a song or play a game.
2 Introduction to workshop: Tell the participants what they'll be able to do as a result of the lesson.

3 Use practical examples of why a map will be useful.
4 Discuss the impact of sun, wind, crops, water sources, floods, droughts, and variable rain on agricultural productivity.
5 Discuss how the map will show relationships between the village, farms, hills, roads hazards, and water sources.
6 Share with the participants that this is their map and that they should do the planning and drawing.

Workshop Participants

Have participants talk about what they do and don't understand, what they do and don't like.

Activity 2: Drawing the Basic Map. Duration 2 Hours

Purpose

To let the participants draw a preliminary draft of the map to work out scale and spatial relationships.

What to Do

1 Take 15 minutes and have community members draw a small preliminary map of the village, surrounding farm fields, roads and pathways, farm fields and watersheds on a single sheet of newsprint. This will accomplish two things: you might discover that there is a good illustrator in the group—and you can quickly solve spatial problems by moving things around. A suggestion is to begin with a black marker to draw the basic outline of the village, roads, pathways, and major farming areas. Later, you can then use different colors for houses, rivers, and farm fields.
2 After the participants have had a chance to discuss the map, encourage them to make corrections to the map.
3 Then, tape several sheets of newsprint together creating a much larger sheet and place it on the floor.
4 Let your group artist transfer this basic outline of the community onto the larger of newsprint.
5 Take colored sheets of paper and cut them out to represent additional features. These could be individual farm fields, houses, schools, and churches—and stick them to the map with removable tape.

Other ideas that could be indicated on the map could include:

- sources of both domestic and agricultural water and their relationship to the village and farmlands;
- seasonal availability of both domestic and agricultural water;
- rivers and streams;
- the location of steep hillsides or canyons;
- community land, forest boundaries, grazing/pasture lands.

Workshop Participants

Have participants talk about what they do and don't understand, what they do and don't like.

Workshop Facilitators Note

1 This map is not meant to be submitted to *National Geographic*; this map is meant to be something that can be readily understood and used by the community members. Their sense of scale and spatial relationships may be different than yours—but if it means that they understand the information better—let it be.
2 A first short workshop could be held for introductions to the topic and for drawing a small, preliminary map. Later, the artist could transfer the information to the larger map and then a second workshop held to add expanded information.

Take a break for 15 minutes.

Activity 3: Expanding the Map with Agricultural Information. Duration 60 Minutes

Purpose

To better understand the interrelationships between their agricultural systems, soil and water.

What to Do

1 When everybody at the workshop is satisfied with the basic map, begin applying useful agricultural information. Examples could be:

   a farmlands vulnerable to drought (or insufficient access to water);
   b farmlands vulnerable to flooding, too much wind exposure, and other weather-related hazards;
   c areas that suffer from excessive runoff;
   d areas of high erosion and gullies;
   e the location of different crops;
   f farmers' perception of the fertility of their soil: good, medium, or poor;
   g types of soil;
   h areas of high and low agricultural production;
   i forest cover and areas of deforestation.

Workshop Facilitators Note

In the next workshop, you will actually visit the farm fields and water sources with an agriculturalist, and upon return will add new information—such as soil quality, water runoff, erosion, and deforestation.

Activity 4: Expanding the Map with Community Hazard Information. Duration 60 Minutes

Purpose

To better understand their interrelationships between agricultural systems, the community, and hazards.

What to Do

1 Next, begin the process of overlaying hazards onto the map. Examples could include:

   a areas of the community prone to flooding—and the source of floodwaters;
   b areas of the community with insufficient access to water;
   c areas of the community most affected by heavy rain or extreme weather events;
   d what areas of the community are the most vulnerable to hazards;
   e which people and which livelihoods are the most vulnerable to hazards.

Activity 5: Conclusion: Open Discussion about the Map and a Summary of Why You Did This. Duration 30 Minutes

Purpose

To discuss and reinforce what has been learned.

What to Do

1 discuss and review what has been learned;
2 reinforce farmers' soil and water challenges that were identified during the mapping exercise;
3 discuss whether hazards such as floods, variable rainfall and drought impact soil and water resources;
4 discuss community vulnerabilities to hazards;
5 discuss whether the intensity of the hazards is increased by farming practices and deforestation.

Workshop Facilitators Note

- What are the community members' current coping strategies for dealing with these difficult challenges?
- Capacity building: Which of the difficult challenges are they having trouble coping with due to a lack of strategies?
- Summarize a list of the challenges and hazards farmers face in preparation for the visit by the extension agent.

FIELD GUIDE 10.6: AGRICULTURAL SOIL AND WATER MANAGEMENT FOR SLOPING LAND

Subsistence farmers suffer not only from depleted soils but also from challenges with water: too little water, too much water, and erosion from water. This field guide looks at different ways of developing barriers on farm fields to stop the flow of water so that it can percolate into the soil and build up soil moisture. The barriers also conserve soil by reducing loss from erosion. Organize a 3-hour workshop with 12–15 farmers from your community.

Barriers to Water Movement

On sloping farm fields, creating barriers reduces the speed of water movement so that it can be absorbed into the soil rather than simply running off the land. These barriers also catch topsoil that the water carries, preventing the loss of this valuable resource, and offer the added benefit of creating level planting areas behind the barriers as the soil accumulates. Barriers can be terraces, stone and earth walls called bunds, or living barriers such as hedges and grass strips.

Building terraces and stone retaining walls can be very labor-intensive. Less formal constructions such as soil bunds, hedgerows or rows of grass can be less labor-intensive and therefore potentially more attractive to farmers.

One thing that all barriers have in common is that they run horizontally along a level contour across the falling slope of a field. An A-frame leveling device is used to determine the level contour lines which are marked with stakes or with stones.

Here are four techniques for farmers to consider. The technique chosen by each individual farmer will be based upon how steeply a farmer's field slopes, how big their field is, whether they are in a high rainfall or low rainfall region, and how much time they have available for investing in the technique.

Contour Ridges

Ridges with furrows on the uphill side are formed approximately 1.5m to 2m apart. This 2m area is the catchment area for rainwater. The ridges are only 15–20 cm high—simply high enough to contain the runoff—which collects in the furrow. Contour ridges represent the least time investment of these four techniques and can be developed, maintained and improved during preparation for each planting season.

Soil Bunds

Soil bunds are a method for both containing water and reducing erosion using on-site materials. After marking the horizontal contour line on the sloping field, a ditch 60 cm deep and 60 cm wide is dug. The soil is placed on the downhill side of the ditch, creating the soil wall. The base of the wall is typically twice as wide as the wall is high. Soil bunds are placed from between 5m apart on steep land to 20m apart on more gently sloping land. To determine spacing between the bunds, one rule of thumb is that the top of one bund is level with the base of the adjacent uphill bund. However, farmer preferences and the size of the farmer's field are other determinants.

The soil should be well compacted by hand, then fodder grasses, trees and crops are planted on the bund to stabilize it. Water collects in the ditch during rainstorms and can slowly percolate into the soil, thus increasing soil moisture. As rainwater erodes soil uphill of the bund, the soil will accumulate above the bund and begin creating an increasingly level planting strip. Soil bunds will need annual maintenance—and will need to be checked after heavy rainfall and breaches repaired immediately.

Hedgerows

Hedgerows can also be planted along the contour lines of a hillside—in similar spacing as soil bunds depending on the steepness of the slope of the field. Hedges are usually chosen from nitrogen-fixing plants, and from plants that when pruned can be used as fodder for farm animals. Initially, these cuttings can be laid at the base of the hedges on the uphill side to trap eroded topsoil. After two or three years, topsoil will begin to accumulate and form a terrace uphill of the hedgerow. Hedgerows represent substantially less time investment than soil bunds—and use less space—making more land available for planting.

Vetiver Grass Strips

An inexpensive alternative, vetiver grass can be planted along the contour line of a sloping field to prevent the loss of topsoil, and to reduce the rate at which water runs downhill, thereby enhancing infiltration. Topsoil builds up on the uphill side and over time creates level planting areas. Grass strips represent substantially less time investment than soil bunds—and use less space. Grass strips need to be maintained over time to keep them from encroaching into the cropping areas. Grass trimmings can be used as fodder. Vetiver grass is very popular, but check with farmers for local favorites.

Conclusion

Upon completion of this workshop, discuss with the farmers which technique would be best suited for them. Then plan a second more specialized training workshop for that specific technique.

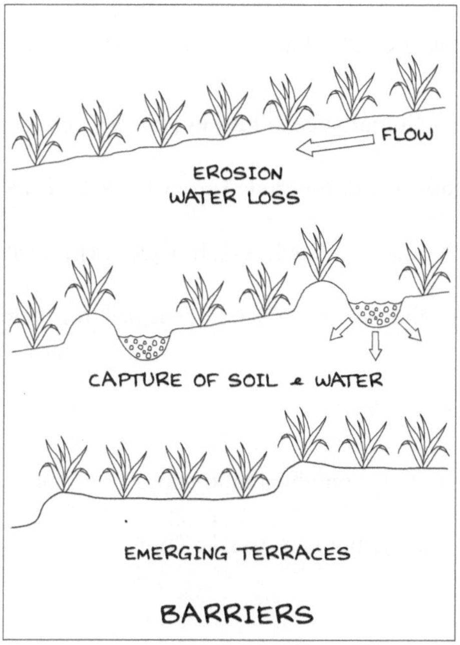

BARRIERS

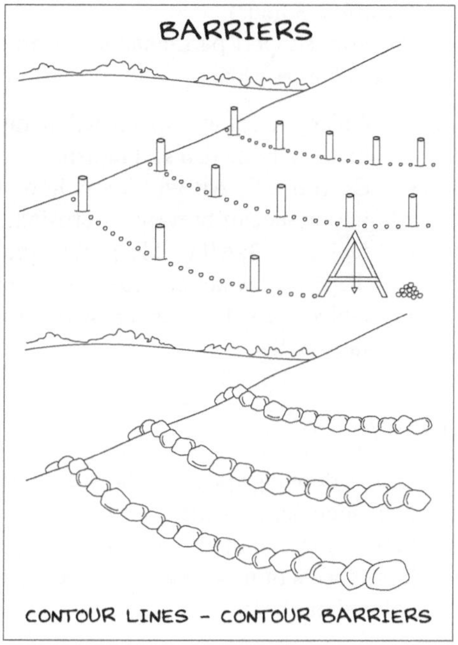

CONTOUR LINES – CONTOUR BARRIERS

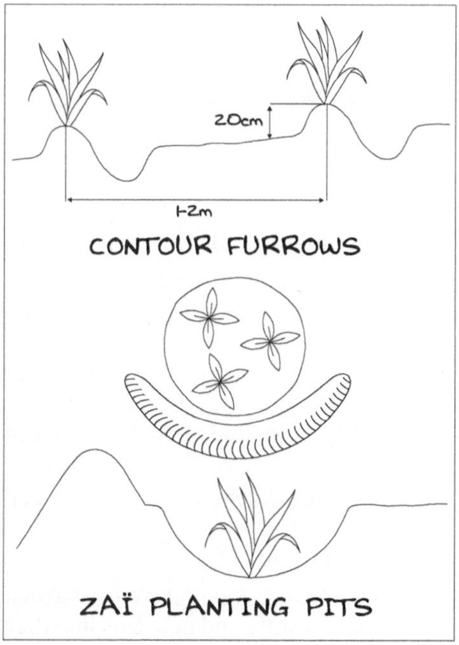

CONTOUR FURROWS

ZAÏ PLANTING PITS

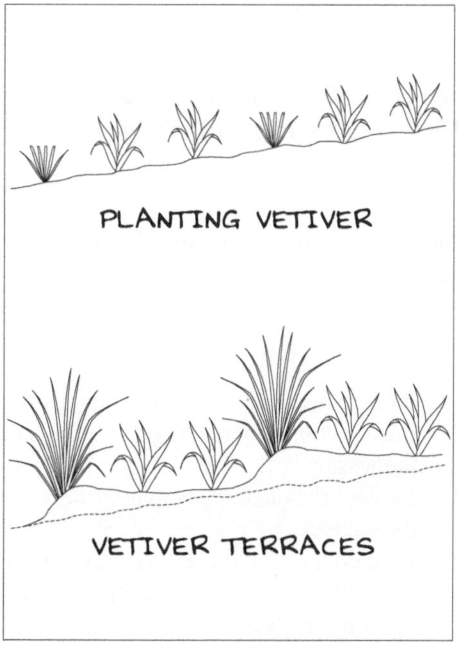

PLANTING VETIVER

VETIVER TERRACES

Workshop Lesson Plan for Agricultural Soil and Water Management for Sloping Land

- *Duration*: 3 hours.
- *Purpose*: support participants in learning about the importance of soil restoration and water conservation:
    - Objective 1: Participants will understand the importance of keeping rainfall on their fields to replenish soil moisture.
    - Objective 2: All will know how much topsoil can be eroded from fields and the importance of preventing erosion.
    - Objective 3: All will learn four simple techniques for barriers on hillsides to prevent the loss of water and topsoil.
    - Objective 4: Participants will learn how to choose which technique is best for their farm fields.

Materials

- artist's drawings/posters; the scenes and people they contain should appear familiar to workshop participants;
- how-to cards without written words for workshop participants to take home;
- large sheets of newsprint and tape;
- colored markers.

Preparation

Find a workshop location.

Activity 1: Soil Restoration and Conservation Introduction. Duration 40 Minutes (Including a 15-Minute Ice Breaker)

Purpose

Participants will understand the importance of keeping rainfall and topsoil on their farm fields.

What to Do

1 Ice breaker: Introductions. Sing a song or play a game.
2 Introduction to workshop: Tell the participants what they'll be able to do as a result of the lesson.
3 Discuss the challenges that farmers face due to a shortage of water and soil moisture on their farms.
4 Discuss the challenges that farmers face due to a loss of topsoil.
5 A practical example of why farmers need to conserve rainwater and prevent the erosion of their farm's topsoil is so that their crops can have the moisture and nutrients that they need for productive harvests
6 Discuss that there are low-cost barriers that can be built to retain both rainwater and topsoil on their farm fields.

Workshop Participants

Have participants talk about what they do and don't understand, what they do and don't like.

Activity 2: What Are Barriers? Duration 20 Minutes

Purpose

To understand the importance of barriers in reducing the loss of rainwater and topsoil.

What to Do

Explain the reasons for using barriers:

1 barriers reduce the speed of water movement so that it can be absorbed into the soil rather than running off the land;
2 barriers catch topsoil that the water carries, preventing the loss of this valuable resource;
3 barriers create level planting areas behind them as the soil accumulates, making sloping fields easier to work in;
4 barriers can be terraces, stone and earth walls called bunds, or living barriers such as hedges and grass strips.

Workshop Participants

Take 5 minutes to talk about what you do and don't understand, what you do and don't like.

Activity 3: Contour Ridges. Duration 20 Minutes

Purpose

To understand what contour ridges do and where and how to build them.

What to Do

1 ridges with furrows on the uphill side are formed approximately 1.5m to 2m apart on gently sloping fields;
2 the ridges are only 15–20 cm high—simply high enough to contain the runoff, which collects in the furrow;
3 contour ridges can be developed, maintained and improved during preparation for each planting season.

Workshop Participants

Take 5 minutes to talk about what you do and don't understand, what you do and don't like.

Activity 4: Soil Bunds. Duration 20 Minutes

Purpose

To understand what soil bunds do and where and how to build them.

What to Do

1 soil bunds are a method for both containing water and reducing erosion using on-site materials;
2 after marking the horizontal contour line on the sloping field, a ditch 60 cm deep and 60 cm wide is dug;
3 the soil is placed on the downhill side of the ditch, creating the wall;
4 soil bunds are placed from between 5m apart on steep land to 20m apart on more gently sloping land;
5 soil bunds should be well compacted by hand;
6 fodder grasses, trees and crops are planted on the bund to stabilize it;
7 water collects in the ditch during rainstorms and can slowly percolate into the soil, increasing soil moisture;
8 soil will accumulate above the bund and begin creating an increasingly level planting strip.

Take a break for 15 minutes.

Activity 5: Hedgerows. Duration 20 Minutes

Purpose

To understand what hedgerows do and where and how to plant them.

What to Do

1 hedgerows can also be planted along the contour lines of a hillside—in similar spacing as soil bunds;
2 hedges are usually chosen from nitrogen-fixing plants, and from plants that can be used as fodder for farm animals;
3 after two or three years, topsoil will begin to accumulate and form a terrace uphill of the hedgerow;
4 hedgerows represent a smaller time investment than soil bunds—and use less space, making more land available for planting.

Workshop Participants

Have participants talk about what they do and don't understand, what they do and don't like.

Activity 6: Vetiver Grass Strips. Duration 20 Minutes

Purpose

To understand what vetiver grass strips do, and where and how to plant them.

What to Do

1 an inexpensive alternative, vetiver grass is planted along the contour line of a sloping field to prevent the loss of topsoil;
2 vetiver grass strips reduce the rate at which water runs downhill, enhancing infiltration;
3 topsoil builds up on the uphill side and over time creates level planting areas;
4 grass strips represent substantially less time investment than soil bunds—and use less space;
5 grass strips need to be maintained over time to keep them from encroaching into the cropping areas;
6 vetiver grass is very popular, but check with farmers for local favorites;

Activity 7: Conclusion. Duration 30 Minutes

Purpose

To reinforce what has been learned and to discuss positive solutions.

What to Do

1 Discuss and review what has been learned.
2 Reinforce the principles of soil restoration and conservation:

    a barriers reduce the speed of water movement so that it can be absorbed into the soil;
    b barriers also catch topsoil that the water carries preventing the loss of this valuable topsoil;
    c barriers create level planting areas behind the barriers as the soil accumulates;
    d barriers can be terraces, stone or earth walls called bunds, or living barriers such as hedges and grass strips.

3 A farmer's choice of barriers will depend on how steep their field is, how big their field is, the level of rain and their available time.
4 Discuss with the farmers what will be the most appropriate barrier for them to build on their farms.
5 Arrange the next workshop with the farmers to begin the training process for building the barriers that they've chosen.

Workshop Participants

Have participants talk about what they do and don't understand, what they do and don't like.

FIELD GUIDE 10.7: HOUSEHOLD ROOFTOP RAINWATER HARVESTING

Introduction

Access to household water has become increasingly challenging in many parts of the world. Frequently women and children must walk several kilometers to collect water—taking time away from livelihood tasks and from school. Rainwater harvesting systems can be easy to build, inexpensive, and provide a meaningful quantity of water. They consist of a roof, a gutter, a downspout and a tank. Organize a workshop for 12–15 community members on rainwater harvesting.

Determine two things prior to offering the workshop. One: that rooftop rainwater harvesting is a viable option for your locale, based upon types of roofing, annual rainfall, length of the dry season, and family water consumption. Two: determine local costs for installations and local material options for use in construction, and investigate water storage system designs that are appropriate for your community context. You should contact a local expert for assistance in designing the systems and in designing the workshop.

The Catchment Surface: The Roof of Your Home

The size of a roof determines how much water can potentially be collected in a year. For example, a 24m² roof in an area with an annual average of 400 mm of annual rainfall can collect and store a three-month supply of water for a family of six. Most roof systems work well for capturing rainwater: the best are concrete, tile or corrugated sheet metal. Thatched roofs can be dirty and difficult to attach a gutter to. However, water collected from thatched roofs can be used in home vegetable gardens. Even 'flat' concrete roofs have a slope and a low point where you can collect water.

Gutters and Downspouts

At the low edge of the roof it is necessary to install a gutter to collect rainwater before it falls to the ground. Gutters can be made of V-shaped pieces of sheet metal, strips of corrugated roofing bent into a C-shape, or PVC or bamboo tubes split in half. These gutters need to be suspended from the roof edge so that all of the rainwater flows into the gutter. The gutter needs to slope downhill from one end to the other end in the direction of your storage tank. Watch during the first few rainstorms to see if your gutter is working effectively. For the downspout: Connect one end of a piece of tubing to the end of the gutter and the other end to an opening in your tank.

The Storage Tank

The tank is the most expensive portion of the system. Because of this, and to simply get started, you can begin collecting rainwater in found containers such as jerry cans or barrels.

The optimal size of the tank will vary, based on how large your roof is, what the annual rainfall is, how long the dry season is, how large your family is, and how much of the captured water is to be stored for future use. Three months of water (a large tank) for a family of six represents half the water they need during a six-month dry season. During unpredictable rainy seasons, a rainwater harvesting system can provide supplemental water during short dry periods with smaller storage containers. Tanks can be aboveground tanks made of concrete blocks, reinforced concrete, or pre-purchased plastic tanks. In-ground cisterns can be as simple as an earth reservoir lined with sheet plastic—or a brick-lined excavated hole.

Planning

Help each participant determine the amount of potential water they can harvest during the course of a year and how much water they need for their family during the dry period. Realizing that costs for building a tank can range from $50 for a plastic lined excavated hole, up to $500 for an aboveground tank of concrete block, work with the participants to determine what their best plan would be for their home and resources. Discuss how you decide which materials to use for the gutter, for the tank—and for the size of the tank for the demonstration installation.

Construction

Deliver the construction materials to the site in advance. Check that you have the correct tools that you need and that you have extras so that several people can participate at the same time. In order to build a system with the time allowed in a workshop, the tank should already have been installed. You should have examples of several types of gutters, downspouts, containers, and plastic sheeting used for lining an excavated hole.

Begin by showing the correct placement of the gutter so that it will effectively catch the water as it flows off of the roof—and how the gutter needs to slope downhill from one end to the other towards the tank. Let the participants install the gutter and the downspout that leads to the tank. Place a piece of wire mesh where the gutter meets the downspout to catch leaves and debris. Run a trial by pouring a bucket of water on the roof to see that the gutter catches the water and that the water flows easily to the tank without leaks.

First Flush

The first rainfall of the season will clean the roof and can potentially flush dirt into the tank. Disconnect the downspout from the tank during the dry season until the first rain can completely clean the roof and the gutters. If the tank is empty—it should also be cleaned out at this time. After the first rain, reconnect the downspout to the tank.

Water Purification

Rainwater collected from rooftops may have impurities in it such as windblown dust or bird droppings. Second, storage conditions may not be optimal. Rodents may get into the storage tank. Water stored for several months may become mildly contaminated. The water needs to be purified or filtered. This could be with point-of-use water filters, with the addition of chlorine, or by using a solar disinfection system.

Maintenance

Roof surfaces need to be kept free from dirt; gutters, drainpipes, and wire mesh must be cleared of leaves and dirt.

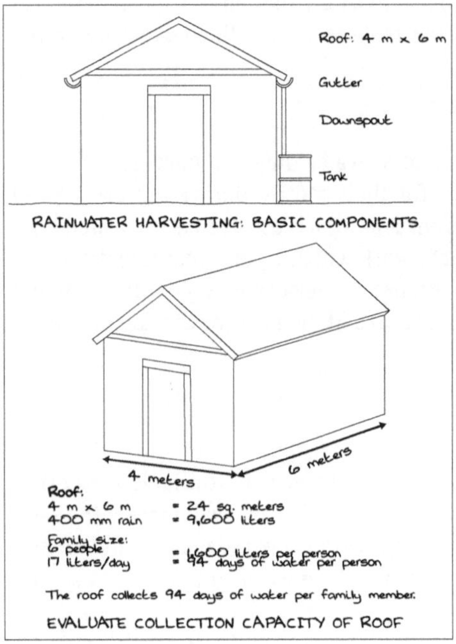

RAINWATER HARVESTING: BASIC COMPONENTS

EVALUATE COLLECTION CAPACITY OF ROOF

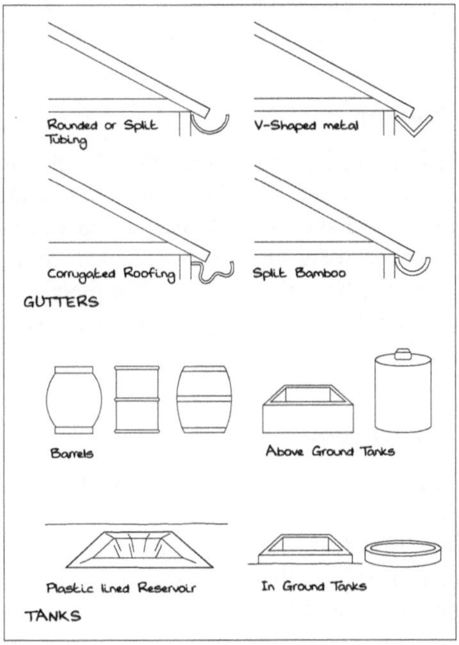

GUTTERS

TANKS

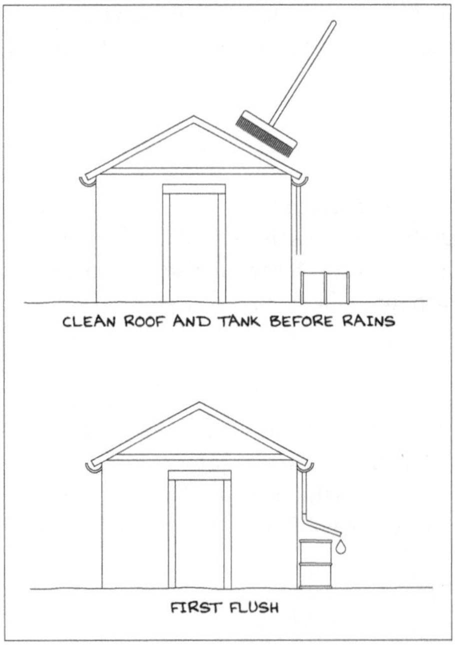
CLEAN ROOF AND TANK BEFORE RAINS

FIRST FLUSH

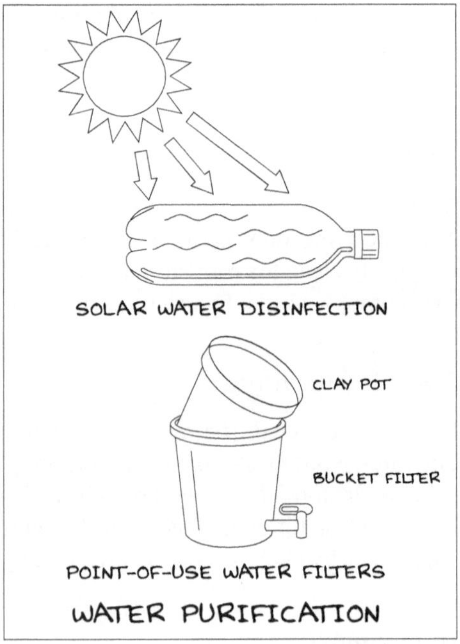

SOLAR WATER DISINFECTION

POINT-OF-USE WATER FILTERS

WATER PURIFICATION

Workshop Lesson Plan for Household Rooftop Rainwater Harvesting

- *Duration*: 5½ hours (can be completed in one 2½-hour workshop and one 3-hour workshop).
- *Purpose*: What workshop participants will be able to do as a result of the lesson:
    - Objective 1: Participants will understand the benefits of rainwater harvesting.
    - Objective 2: All will understand the basic design components and appropriate materials for use in construction.
    - Objective 3: Participants will know how much water they can collect in a year and how much their family needs.
    - Objective 4: Participants will know how to assemble and maintain a rooftop rainwater harvesting system.

Materials

- artist's drawings/posters; the scenes and people they contain should appear familiar to workshop participants;
- how-to cards without written words for workshop participants to take home;
- large sheets of newsprint and tape;
- colored markers;
- example materials to be used for gutters, downspouts, containers, and storage tanks;
- tools for installing the gutter, downspouts and, if appropriate, the storage tank.

Preparation

- Determine that rainwater harvesting is viable in your location. Determine costs and appropriate local materials.
- Find a home that belongs to a member of the community willing to host a demonstration installation.
- Materials for installation should be on site prior to the workshop. The tank should have been installed.

Activity 1: Introduction to the Benefits of Rooftop Rainwater Harvesting. Duration 40 Minutes (Including a 15-Minute Ice Breaker)

Purpose

Participants will understand the benefits of rainwater harvesting.

What to Do

1 Ice breaker: Introductions. Sing a song or play a game.
2 Introduction to workshop: Tell the participants what they'll be able to do as a result of the lesson.
3 Participants will learn that a rainwater harvesting system can provide a meaningful quantity of water for their families.
4 They will learn a system consists of a roof, a gutter, a downspout and a tank.

5 All will learn much of the system can be inexpensive, and that expensive components (the tank) can be phased in over time.

Workshop Participants

Take 5 minutes to talk about what you do and don't understand, what you do and don't like.

Activity 2: The Catchment Surface: The Roof of Your Home. Duration 15 Minutes

Purpose

To understand how the roof functions in the system.

What to Do

1 discuss how a roof catches rainwater that flows to the low edge of the roof and can be collected in a gutter;
2 discuss that concrete, tile, or corrugated sheet metal perform best;
3 thatched roofs can be dirty, can be difficult to attach gutters to, but can collect rainwater for vegetable gardens;
4 prior to the rainy season, the roof should be cleaned and should be kept free of leaves during the rainy season.

Workshop Participants

Have participants talk about what they do and don't understand, what they do and don't like.

Activity 3: The Gutter and Downspout. Duration 15 Minutes

Purpose

To understand how the gutter and downspout function in the system and how they are installed.

What to Do

1 show samples of what a gutter looks like and samples of the different materials that they can be made from;
2 show the position where the gutter should be placed, so that it can catch all of the rainwater flowing off of the roof;
3 show how the gutter needs to slope downhill from one end to the other towards the tank;
4 show how the downspout tubing connects to the end of the gutter to direct the water into the tank.

Workshop Participants

Have participants talk about what they do and don't understand, what they do and don't like.

Activity 4: The Tank. Duration 20 Minutes

Purpose

To understand what alternative designs there are for tanks and what materials can be used.

What to Do

1 discuss the fact that tanks can be an investment. Other containers such as barrels can be used temporarily;
2 describe how a small reservoir can be dug in the ground and lined with plastic to store water;
3 describe different styles of aboveground tanks and the different materials that they can be made from;
4 describe different styles of underground tanks and the different materials that they can be made from;
5 discuss the different materials' costs and discuss the different benefits that each tank style offers;
6 the size of a tank will depend on the size of the roof, the annual rainfall, the length of the dry season, and the size of the family.

Take a break for 15 minutes.

Activity 5: Planning. Duration 20 Minutes

Purpose

For each person to experiment with the best design options and the best investment level for their family.

What to Do

1 each participant will determine how much water they can capture, and how much their family needs per year;
2 since tank costs vary substantially, help the participants determine which option is best for them;
3 discuss how you decided on the materials to use for the gutter, tank—and for the size of tank for the demonstration.

Workshop Participants

Have participants talk about what they do and don't understand, what they do and don't like.

Activity 6: Construction. Duration 2 Hours

Purpose

To learn the steps involved in installing a rainwater harvesting system.

What to Do

1 reinforce the correct placement and downhill slope of the gutter and let the participants install the gutter;
2 show how to connect the downspout/tubing which leads to the opening in the tank—and let the participants install it;
3 place a piece of wire mesh over the opening where the gutter meets the downspout to catch leaves and debris;
4 test the installation by pouring a bucket of water on the roof. See that the gutter catches the water and there are no leaks.

Activity 7: First Flush, Use, and Water Purification. Duration 20 Minutes

Purpose

To learn the steps in maintaining a clean system and purifying the water before using it.

What to Do

1 the first rainfall will clean the roof and can flush dirt into the tank;
2 disconnect the downspout from the tank during the dry season until after the first rain cleans the roof and the gutters;
3 clean the tank out (directions below) at the end of the dry season and before the beginning of the rains;
4 underground tanks may require a hand pump or bucket to retrieve water. Aboveground tanks can use spigots;
5 rainwater collected from rooftops may have impurities in it such as windblown dust or bird droppings;
6 drinking water needs to be purified or filtered with point-of-use water filters, chlorine, or by using a SODIS.

Activity 8: Maintenance. Duration 20 Minutes

Purpose

To learn the steps involved in maintaining and repairing a rainwater harvesting system.

What to Do

1 Regular maintenance (dry season):

   a roof surfaces have to be kept free from bird droppings and other dirt;
   b gutters, drain pipes, and inflow filters must be regularly cleared of leaves and dirt;
   c disconnect the downspout from the entrance into the tank at end of the dry season until after the first rain for cleansing flush.

2 Annual maintenance (at the end of the dry season):

   a at the end of the dry season, when the tank is empty, repair leaks;
   b repair roof, gutters, inflow pipes, filters;
   c remove deposits from bottom of tank;
   d mix 1 cup of chlorine bleach (5%) with 45 liters of water, pour into tank, scrub clean with chlorine solution, leave tank with solution in it for 24 hours. Flush tank out with clean water—then connect downspout to clean tank.

Activity 9: Conclusion. Duration 30 Minutes

Purpose

To review rainwater harvesting system and provide feedback.

What to Do

1 reinforce the benefits of rainwater harvesting;
2 review the basic design components and appropriate materials to use for construction;
3 review how much water can be collected at a participants' home and how much water is needed by the family;
4 review the planning process of designing the system and the costs of installing a system;
5 review the first flush, use of the system, and maintenance of the system.

Workshop Participants

Take 10 minutes to talk about what you do and don't understand, what you do and don't like.

Follow-up

Your organization should set up a follow-up schedule with each of the workshop participants to assist them in the installation of their harvesting systems.

FIELD GUIDE 10.8: COMMUNITY-LEVEL WATER HARVESTING

Developing Community Water Harvesting as a Component of a Master Water Use Management Plan

A shortage of water or unreliable access to water is one of the biggest issues in development. Community water sources dry up during climate change-related drought—or seasonally during the dry season. There is competition among different segments of the community for available water. Discussing community water harvesting calls for stepping back from the immediate problem and looking at the relevant underlying causes for the shortage of water in order to begin developing solutions.

Community Based Water Management Committees

Forming a water management committee is the initial step in developing a water management plan. A committee can provide direction, consistency, management, regulation and enforcement.

The first step in forming a water management committee is to introduce the concept in a participatory workshop setting; ask your community contacts to approach community members about participating in a water management workshop. In the workshop, you can begin by introducing the rationale and importance of having a community based water management committee who can develop a plan for improving access to water.

Discuss the types of skills that will need to be developed by committee members. Suggest appointing an interim committee who over the course of six months could develop the ultimate committee, establish management responsibilities, ensure gender inclusiveness, set goals, and initiate the planning process. Ask the workshop participants to decide on the number of people that should serve on the committee—and then ask if there are members who would be interested in sitting on the interim committee.

Water Use Management Plan

The first step in developing a water use management plan is to better understand what knowledge community members have about water resources and water use. A simple way to do this is to hold a participatory mapping workshop to look at the community's water resources, distribution, uses, and weather related challenges. Field Guide 5 explains how to do this first step.

Having identified the community's sources of water, how the water is distributed, how the water is used, when there's too much water and when there's too little water, you can begin the process of developing a plan which would include:

1. *Consulting with a water management expert to develop a participatory training and planning program.* A water management professional will be able to help the community assess their unique situation and propose solution-oriented activities appropriate for community context, capabilities and resources.
2. *Prioritizing which challenges to approach first.* The challenge of not enough water could be due to unequal allocation, deforested catchment areas, existing infrastructure that is in disrepair, or it could be that new, alternative sources have not been explored.

In the community's existing context, determine which challenge, if solved, will give them the greatest increase in water access at the lowest cost.

3 *An investigation into the restoration of environmental services.* If the watershed around the village has been deforested over time, it is possible that rainwater simply runs off the hillsides rather than infiltrating into the soil and recharging the local water table. The water management expert can help with this determination and help develop a plan for reforestation.

4 *An investigation of water sourcing alternatives.* The best alternative may be to look for new sources of water. This could include a new well, or it could include water harvesting structures to channel water to a community storage reservoir.

5 *The development of an infrastructure installation, maintenance, and repair program.* Two of the main reasons why water systems fail over time are that they were not designed by experts and they were not maintained by the community. A sustainable water use management plan needs to take into consideration expertise in both design and installation oversight, and a plan for funding and implementing routine maintenance.

6 *The development of a water allocation plan.* The community may have sufficient water but a wealthy farmer or nearby factory may be using more than their share. Developing an advocacy plan to approach the local government can begin the process of creating an allocation plan which is fairer to the community. An allocation plan will also ensure that new community water resources are distributed fairly among different segments of the community.

7 *Compile these steps into a comprehensive water use management plan.* Completing these steps could take a year or more but when they are complete, they need to be compiled into a master water use management plan. The water management committee should be in charge of overseeing the development of the plan's components, of collecting water use fees, and of enforcing the plan.

The Water Challenge

Here is an example project detailing water challenges and solutions that are typical in rural settings. During the course of their investigations, a water use management committee determined that deforestation of the hilly watershed behind the village has led to rainwater runoff that causes soil erosion, the creation of gullies, siltation of the main stream bed, and occasional flooding in the village during heavy rain. The downhill movement of rainwater is no longer slowed by trees and undergrowth; it no longer slowly infiltrates into the soil and recharges the local groundwater system. This is evident in the fact that the village well and the streambed are frequently dry.

The Water Management Plan

The committee decided upon a triple plan: a soil and water conservation program in the watershed, a reforestation program for the watershed, and the construction of a sub-surface dam in the dry streambed adjacent to the village well.

With the help of a watershed management expert, a more detailed map was drawn of the village's watershed showing where rivulets and gullies direct rapidly flowing rain water away from the watershed. The expert suggested locations for 12 check dams on these gullies and said they should be the first step. Their purpose is to slow the velocity of water movement, thereby reducing erosion and allowing water to infiltrate into the soil.

Water infiltration will recharge local groundwater systems and provide soil moisture for the reforestation program. A reduction in runoff from heavy rains will reduce flooding in the village. Over time, soil accumulating behind the check dams will begin the process of filling in the gullies.

The plan is to start a tree seedling nursery of native trees, and at the same time begin the construction of the 12 check dams. The villagers are concerned about building the sub-surface dam near the village until the check dams have been put in place; they don't want the sub-surface dam to wash away during a flood if the check dams are ineffective. They have time now to build the check dams before the next rainy season—and they will have tree seedlings to plant alongside the gullies and rivulets when the rainy season begins. If their check dams are successful and the village doesn't flood, then they will feel comfortable building the sub-surface dam in the riverbed at the beginning of the next dry season.

Because of the expense and engineering requirements of concrete check dams, the community decided to experiment with very simple, inexpensive check dams to ensure that they were going to be effective. Two kinds of check dams were chosen: loose stone and brushwood. They are semi-porous: some water will pass through them and reduce water pressure build-up. Consequently the engineering requirements are minimal. Rainfall conditions, terrain and building materials vary from location to location, and the check dam expert helped the community members decide the size and configuration of their dams. Three workshops have been planned. A one-day workshop to build a brushwood check dam, and a two-day workshop to build a loose stone check dam. These check dams will be made of locally found materials. If effective, these low-cost check dams could be replaced over time with more permanent structures.

Brushwood Check Dams

A brushwood check dam is a framework of vertical posts sunk 1 meter into the ground 0.5 meters apart. Two parallel rows of these posts are placed 0.5 meters apart. These types of dams are rarely more than 1 meter high and the catchment area should be 1 hectare or less. The first step is to clear the dam's location in the gully of brush and loose soil. Cut a slot 30 cm deep into the walls of the gully. Dig a series of 15 cm diameter holes for the posts, a meter deep, with a post-hole digger. Ram the posts vertically into their holes and pack dirt around them so they are rigid. Cut brush or branches 2 or 3 cm in diameter into 65 cm lengths. Tie them into bundles 15 cm in diameter. Do not place these bundles directly on the ground, but on a bed of straw or mulch to prevent erosion under the brushwood. Place them in layers inside of the framework making sure each layer is well compacted. Every second layer, cross-stays should be tied to the vertical posts connecting the two rows of posts. The bundles should penetrate the walls into the 30 cm deep slot and anchor the dam. The top of the check dam should be curved with a low section in the center for water to spill over if the check dam fills up. Bundles of sticks should cover the ground on the downhill side of the dam so that soil below the dam won't be eroded.

Loose Stone Check Dams

Typically, these types of dams are rarely more than 1 meter high and the catchment area should be 2 hectares or less. The first step is to clear the dam's location in the gully of brush and loose soil. The second step is to dig a 30 cm deep trench beginning at just below the top lip of one bank, progressing across the bottom of the gully and extending to just below the top lip of the opposite bank. This will allow the stone dam to be anchored in position. The highest point of the dam should be lower than the surrounding land to prevent flooding of adjacent fields. The base of the dam should be wider than the top—so the walls slope inwards from the base towards the top. The first layer of stones in the trench acts as the anchor; this layer of anchor stones will be one stone width wide. The next layer of stones should be three stones wide and after several layers narrow to being two stones wide as it rises to the top. The top of the check dam should be curved with a low section in the center for water to spill over if the check dam fills up. A layer of stones should cover the ground on the downhill side of the check dam so that soil below the dam won't be eroded as water spills over the top.

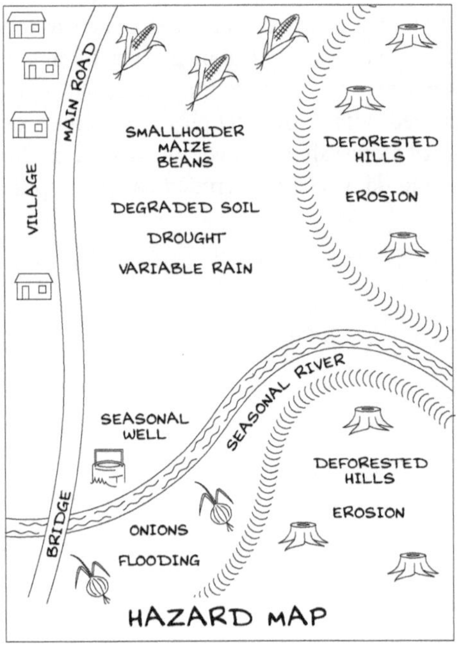

VILLAGE
MAIN ROAD
SMALLHOLDER MAIZE BEANS
DEGRADED SOIL
DROUGHT
VARIABLE RAIN
DEFORESTED HILLS
EROSION
SEASONAL RIVER
SEASONAL WELL
BRIDGE
ONIONS
FLOODING
DEFORESTED HILLS
EROSION
HAZARD MAP

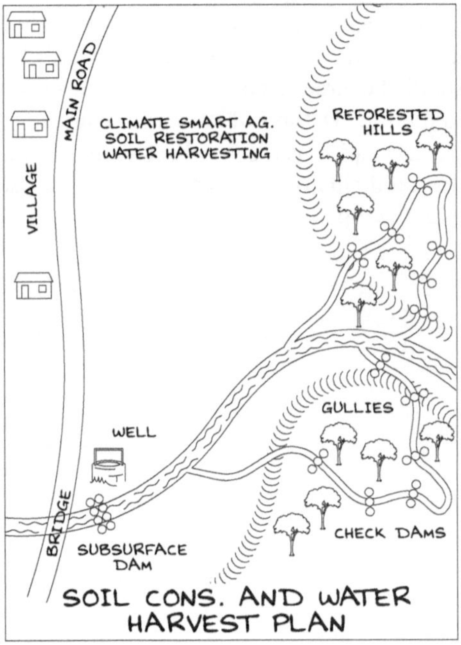

VILLAGE
MAIN ROAD
CLIMATE SMART AG.
SOIL RESTORATION
WATER HARVESTING
REFORESTED HILLS
GULLIES
WELL
BRIDGE
SUBSURFACE DAM
CHECK DAMS
SOIL CONS. AND WATER HARVEST PLAN

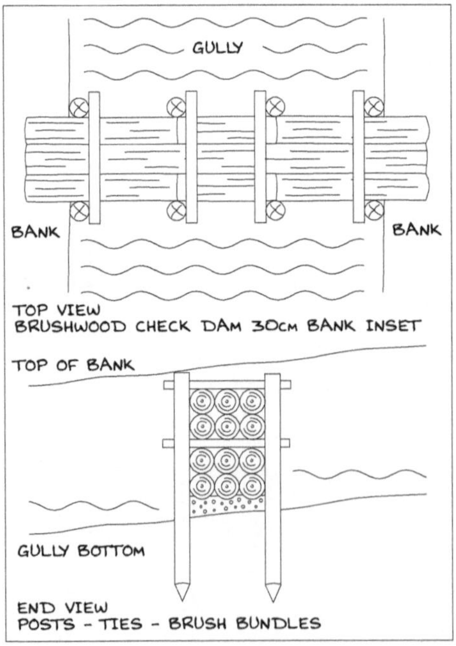

GULLY
BANK
BANK
TOP VIEW
BRUSHWOOD CHECK DAM 30CM BANK INSET
TOP OF BANK
GULLY BOTTOM
END VIEW
POSTS - TIES - BRUSH BUNDLES

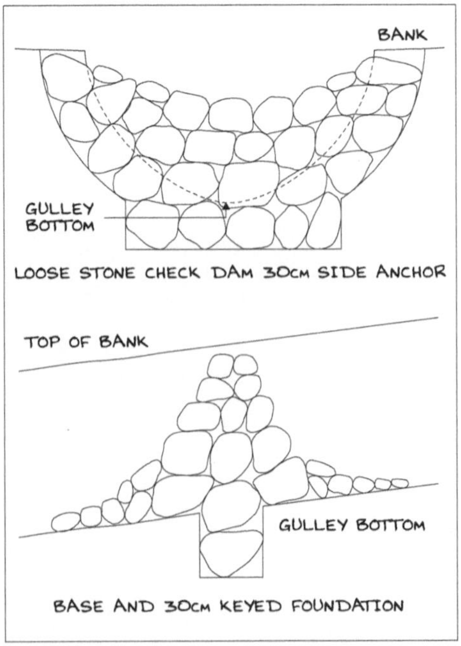

BANK
GULLEY BOTTOM
LOOSE STONE CHECK DAM 30CM SIDE ANCHOR
TOP OF BANK
GULLEY BOTTOM
BASE AND 30CM KEYED FOUNDATION

Workshop Lesson Plan for How to Build Check Dams

- *Duration*: Three days (can be completed in three one-day workshops or six half-day workshops).
- *Purpose*: What workshop participants will be able to do as a result of the lesson:
  - Objective 1: Participants will understand the benefits of a building check dams.
  - Objective 2: All will understand the basic design components and appropriate materials for use in construction.
  - Objective 3: Participants will know how check dams can reduce loss of topsoil and recharge groundwater.
  - Objective 4: Participants will know how to construct and maintain two kinds of check dams.

Materials

- artist's drawings/posters; the scenes and people they contain should appear familiar to workshop participants
- how-to cards without written words for workshop participants to take home
- large sheets of newsprint and tape
- colored markers
- tools for installing the check dams including picks, shovels, sledgehammer, brushknife, and galvanized wire
- stones for the loose stone check dam, and posts and brush for the brushwood check dam

Preparation

- identify a gully in need of restoration close to the village;
- deliver sufficient stones for building the stone check dam to the site before the workshop;
- deliver posts and sufficient cut brush for the brushwood check dam to the site before the workshop.

Activity 1: Introduction to the Benefits of Check Dams. Duration 40 minutes (Including a 15-Minute Ice Breaker)

Purpose

Participants will understand the challenges of deforestation and opportunities for restoration.

What to Do

1. Ice breaker: Introductions. Sing a song or play a game.
2. Introduction to workshop: Tell the participants what they'll be able to do as a result of the lesson.
3. All will learn deforested hillsides (the watershed) can lead to topsoil erosion, gullies, flooding, and a loss of groundwater.
4. They will learn the loss of groundwater can mean that streams and wells dry up.
5. All will learn that watersheds can be restored to reduce runoff using several systems—one of which are check dams.

Workshop Participants

Have participants talk about what they do and don't understand, what they do and don't like.

Activity 2: What Check Dams Do. Duration 15 Minutes

Purpose

To understand how check dams work.

What to Do

1 discuss how a low wall placed across a rivulet or a gully will slow down the movement of water during heavy rain;
2 this allows water to infiltrate into the soil—recharging groundwater that charges their wells;
3 slowing runoff allows eroded topsoil to collect behind the dam, restoring the landscape;
4 check dams can be solid concrete dams—or semi-porous dams that allow some water to pass through;
5 porous dams such as loose stone dams and brushwood dams are less expensive to build and require less engineering.

Workshop Participants

Have participants talk about what they do and don't understand, what they do and don't like.

Activity 3: Building a Brushwood Check Dam. Duration One Day

Purpose

To understand how to build a brushwood dam.

What to Do

1 show the participants how the site for building the dam was chosen;
2 have one team begin digging a slot 30 cm deep by 50 cm wide into the walls of the gully;
3 have another team begin digging 15 cm diameter holes for the posts a meter deep in two rows 50 cm apart;
4 the 15 cm diameter posts should be cut to be approximately 2 m long and then pointed at one end;
5 have one team begin pounding the posts into the holes with a sledgehammer and then tightly packing soil around them so that they are vertical and rigid;
6 have a second team take the cut brush and cut off the twigs leaving 2 or 3 cm diameter sticks 65 cm long;
7 make 15 cm diameter bundles and tie the bundles with sisal or wire;
8 lay straw or other dry mulching materials on the ground at the bottom of the gully between the two rows of posts;

9 begin putting the bundles in between the parallel posts and compressing them in place, making sure the ends of the bundles are inserted into the 30 cm deep slots in the gully walls in order to anchor the dam in place;
10 on top of every second layer cross-stays should be tied to the posts to structurally connect the two rows of posts;
11 the top of the check dam should be curved with a low section the center for water to spill over;
12 bundles of sticks should cover the ground below the dam so the soil at the bottom of the dam won't erode.

Workshop Participants

Have participants talk about what they do and don't understand, what they do and don't like.

Activity 4: Maintenance. Duration 20 Minutes

Purpose

To learn the steps involved in maintaining and repairing a check dam.

What to Do

1 the dam should be checked before the beginning of the rainy season to make sure that there aren't any gaps;
2 posts and sticks that have rotted should be repaired or replaced;
3 the area below the dam should be checked to make sure that there are sticks on the ground to prevent erosion.

Activity 5: Conclusion. Duration 30 Minutes

Purpose

To review the brushwood check dam system and provide feedback.

What to Do

1 reinforce the benefits of check dams;
2 review the basic design components and appropriate materials to use for construction;
3 review how check dams can reduce loss of topsoil and recharge groundwater;
4 verify that workshop participants are comfortable that they can build a brushwood check dam.

Workshop Participants

Have participants talk about what they do and don't understand, what they do and don't like.

Activity 6: Building a Loose Stone Check Dam. Duration Two Days

Purpose

To understand how to build a loose stone check dam.

What to Do

1. show the participants how the site for building the dam was chosen;
2. the highest point of the dam should be lower than surrounding land to prevent flooding of adjacent fields;
3. have one team begin digging a 30 cm deep by 40 cm wide anchor trench starting just below the lip of one of the walls of the gully, continuing along the ground and then continuing to just below the lip of the opposite wall of the gully;
4. the trench will anchor the check dam into position;
5. the base of the dam should be wider than the top so the walls slope in from the base towards the top;
6. lay a single row of stones in the foundation trench;
7. then lay three rows of stone above the foundation stones and begin building the wall up;
8. the wall should slowly become narrower as it rises and only be two stones wide at the top;
9. the top of the check dam should be curved with a low section in the center for water to spill over;
10. more stones should cover the ground below the dam so the soil at the bottom of the dam won't erode.

Workshop Participants

Have participants talk about what they do and don't understand, what they do and don't like.

Activity 7: Maintenance. Duration 20 Minutes

Purpose

To learn the steps involved in maintaining and repairing a check dam.

What to Do

1. the dam should be checked before the beginning of the rainy season to make sure that there are no gaps;
2. missing stones should be replaced;
3. the area below the dam should be checked to make sure that there are stones on the ground to prevent erosion.

Activity 8: Conclusion. Duration 30 Minutes

Purpose

To review the loose stone check dam system and provide feedback.

What to Do

1 reinforce the benefits of check dams;
2 review the basic design components and appropriate materials to use for construction;
3 review how check dams can reduce loss of topsoil and recharge groundwater;
4 verify that workshop participants are comfortable that they can build a loose stone check dam.

Workshop Participants

Have participants talk about what they do and don't understand, what they do and don't like.

FIELD GUIDE 10.9: OVERVIEW OF DEVELOPING A COMMUNITY BASED DISASTER RISK REDUCTION PLAN

It is estimated that over 50 percent of all disasters are now related to extreme weather events. Because of this, disaster risk reduction should become an integral part of adaptation projects. Community based disaster risk reduction (CBDRR) has the same merit that community based adaptation does: ownership and sustainability. This field guide presents an overview of establishing a CBDRR program in a community.

Conducting a Participatory Capacity and Vulnerability Analysis

In Chapter 2 you conducted a participatory capacity and vulnerability analysis as part of developing your project. Use the same resources (Chapter 10, Field Guide 2) yet focus the assessment on hazards and disasters. Examples could be floods or extreme weather events such as hurricanes. One of the exercises in Field Guide 2 is to draw a participatory map. It is useful to transfer the information from the map into a larger format and place it on a public wall where everyone in the community can see it and better understand how disasters can impact their village.

Setting up a CBDRR Committee

In Chapter 7, you set up a community based project management committee. You can use the same technique to set up a community based DRR committee. This committee will be able to work with your NGO to create long-term associations with government agencies that can support long-term DRR activities. The committee will be in charge of developing a DRR plan, and for coordinating disaster teams.

Developing a DRR Plan

Developing a DRR plan will include each of the activities in the following list—and maintaining them in perpetuity. This will involve a plan for consciousness-raising among community members about DRR challenges, connecting with an early warning system, organizing teams, training teams in evacuation and search and rescue, and prioritizing mitigation strategies.

Organizing Teams

Based upon the results of your community's participatory capacity and vulnerability assessment, you should now have prioritized a list of hazards and disasters. As part of the DRR plan, the committee will prioritize preparedness activities, risk reduction activities and mitigation activities. Teams should be organized for each of the major priorities. If, for example, the type of disaster your community faces necessitates evacuation, an evacuation team should be established that develops a plan to lead an evacuation at the appropriate time—and then lead an actual evacuation if necessary.

Promotion to Community

Frequently community members don't have a clear picture of how and why disasters happen. They also may not know how to react when a disaster is building or is already in progress.

Workshops and simple posters or how-to cards need to be developed to help them understand these concepts and to learn that there are things that they can do to reduce the risk caused by disasters, and to mitigate the severity of the disasters.

Early Warning Systems

Traditionally, community members have not had warning of when they need to evacuate—and frequently they have left it too late. Your government's meteorological office may have the capability of, for example, in a potential flood situation, evaluating when water has reached a critical height and can announce that a flood is imminent. Communities should form partnerships with these offices and ensure that they have clear access to early warning information in case of disaster.

Evacuation Training

If people do need to evacuate, they need to know when to evacuate, they need to know what to do with their valuable possessions and assets, they need to know where to go where it's safe, they need to know what to take with them, and they need to know what to do when they get to shelter. Capacity building workshops led by the evacuation team can train community members in each of these—and most importantly—can lead them in practice drills.

Search and Rescue Training

Search and rescue team members learn specific techniques that are safe, and are given simple tools such as lifejackets, safe boats, inner tubes, and flashlights which give them the confidence and the capability to look for a missing person or to rescue a trapped elderly or disabled person.

Capacity Building for Disaster Mitigation Activities

There are many activities that can be done to mitigate potential disasters. Some, like mangrove plantings along river banks can reduce erosion during a flood season. The reforestation of a watershed can reduce danger from flash floods. Both are activities that communities can do over time. Capacity building workshops give community members the skill sets that they need to do these activities. Other activities like relocating houses to higher ground may be outside of their purchasing power, but committee members can be trained to develop advocacy campaigns for approaching governments for support for more major investments.

Workshop Lesson Plan for Introducing Community Based DRR Planning: A Workshop for DRR Committee Members

- *Duration*: six hours plus lunch (can be completed in two three-hour workshops).
- *Purpose*: What workshop participants will be able to do as a result of the lesson:
  - Objective 1: Participants will understand the importance of having a disaster risk reduction plan.
  - Objective 2: Disasters, risks associated with them, community vulnerability and capacity will be clearly spelled out.
  - Objective 3: The need for mitigation and potential mitigation activities will be clearly spelled out.
  - Objective 4: Important elements for the plan and a framework for developing the plan will be carefully spelled out.
  - Objective 5: DRR teams will be defined and team coordinators elected.

Preparation

- A community based DRR committee has been formed.
- A participatory capacity and vulnerability assessment has recently been completed.

Materials

- artist's drawings/posters; the scenes and people they contain should appear familiar to workshop participants;
- large sheets of newsprint and tape;
- colored markers.

Activity 1: Introduction to the Benefits of Having a DRR Plan. Duration 1 Hour

Purpose

Committee members will understand the benefits of having a DRR plan.

What to Do

1 Introduction to workshop: Tell the participants what they'll be able to do as a result of the lesson.
2 Participants will learn that community members are frequently unaware of how and why disasters originate.
3 They will learn that community members are not always aware of what to do before, during, and after a disaster.
4 They will learn that there are techniques for reducing the risk caused by disasters.
5 They will learn that a plan can be developed to help community members reduce each area of risk.
6 DRR teams need to be developed and will be assigned to key individuals (coordinators) for formation.

7 Let them know that a DRR professional will accompany them on the journey of developing the plan
8 Assign one committee member to be responsible for the overall production of the plan.
9 Discuss how the plan may take several months to develop.

Workshop Participants

Have participants talk about what they do and don't understand, what they do and don't like.

Activity 2: Prioritizing the Results of the Participatory Capacity and Vulnerability Assessment. Duration 1 Hour

Purpose

To review, prioritize, and clearly state the results of the recently completed PCVA.

What to Do

1 clearly note what disasters the community suffers from or may potentially suffer from;
2 clearly note where (location) the community is most vulnerable and which community members are most vulnerable;
3 clearly note what livelihood assets and resources are most vulnerable;
4 note what the community's knowledge of disasters and the risks are;
5 make a prioritized list of the disasters, vulnerable locations in the community, and livelihood assets and lives most at risk.

Workshop Participants

Have participants talk about what they do and don't understand, what they do and don't like.
Have a break for 15 minutes.

Activity 3: Introduction to Consciousness-Raising within the Community. Duration 30 Minutes

Purpose

For committee members to better understand the need for consciousness-raising and some of the challenges.

What to Do

1 review what the community members do and do not know about disasters;
2 discuss how some community members may have strong biases of what to do in a disaster;
3 discuss how some community members are illiterate and so posters and handouts should not have written words;

4 describe different ways of getting information out to the community to increase capacity and develop responsibility;
5 discuss how important schools and school children can be in disseminating this information;
6 discuss how after initial consciousness-raising, teams will need to provide capacity building to the community.

Take a lunch break for 30 minutes. If the workshop is to be held in two sessions, this is a good breaking point.

Activity 4: Introduction to Early Warning Systems. Duration 30 Minutes

Purpose

For participants to learn what an early warning system is and how important it is:

1 Participants will learn that frequently meteorological offices can provide early warning information to communities.
2 Tell them how early warning systems work.
3 The early warning team will need to learn how to connect to early warning system networks.
4 The early warning team may need to purchase a dedicated telephone and an alarm system.
5 Your NGO will help develop a training program for the early warning team.

Activity 5: Introduction to Evacuation. Duration 30 Minutes

Purpose

To learn the steps involved in evacuating vulnerable community members.

What to Do

1 describe how the evacuation team will need to determine a safe place to evacuate to;
2 describe how the shelter may need to have basic provisions;
3 describe how the evacuation team needs to interact with the early warning team;
4 describe what community members should do with valuable assets, what they should take with them, and what they should leave behind during an evacuation;
5 describe how the evacuation team needs initial and ongoing training;
6 they may need to buy basic equipment like megaphones and flashlights;
7 describe how the evacuation team needs to develop an evacuation plan—and practice and rehearse;
8 describe how community members need to practice and rehearse the evacuation procedures with the team;
9 your NGO will help develop a training program for the early warning team.

Activity 6: Introduction to Search and Rescue. Duration 30 Minutes

Purpose

To have a simple introduction to search and rescue during disasters.

What to Do

1 describe the basics of a search and rescue operation during a disaster;
2 describe how the search and rescue team needs to interact with the early warning team and the evacuation team;
3 describe how a search and rescue team needs initial and ongoing training;
4 they may need to buy basic equipment such as life jackets, inner tubes, stretchers, a boat, and flashlights;
5 describe how the search and rescue team needs to develop a search and rescue plan—and practice and rehearse;
6 the search and rescue team will need to hold consciousness-raising workshops with community members;
7 your NGO will help develop a training program for the search and rescue team.

Workshop Participants

Have participants talk about what they do and don't understand, what they do and don't like.

Have a break for 15 minutes.

Activity 7: Introduction to Disaster Mitigation Activities. Duration 30 Minutes

Purpose

To learn about basic mitigation activities and levels of investment of labor and money.

What to Do

1 discuss with the members simple mitigation activities that can be done to reduce risk for their specific type of disaster;
2 discuss with the members that the mitigation team will need to develop a plan for activities and for funding;
3 discuss with the members that community members will need to be trained in implementation of the activities;
4 your NGO will help in developing a mitigation plan and training program.

Activity 8: Organizing Teams. Duration 1 Hour

Purpose

To initiate teams for reducing disaster risk in the community.

What to Do

1 Review the prioritized list of the disasters, vulnerable locations in the community, and livelihood assets and lives at risk.
2 Make a list of teams that need to be developed to reduce risk within this prioritized list.
3 Teams may include:

a a training team for consciousness raising and capacity building within the community;
b an early warning team;
c an evacuation team;
d a search and rescue team;
e a mitigation team.

4 Discuss who would be the best people in the committee to coordinate the formation of each team.
5 Elect individual team coordinators.
6 Discuss with the coordinators who would be the most qualified community members to lead the individual teams.

Activity 9: Conclusion. Duration 30 Minutes

Purpose

To review the elements of developing a DRR plan.

What to Do

1 reinforce the benefits of developing a DRR plan. Review the basic components of the plan;
2 review who is in charge of which components (the writing of the plan and the formation of the teams);
3 set a goal for a timeframe of completing the plan;
4 set a goal for a timeframe of establishing and training the teams;
5 your NGO will help in developing the plan.

Workshop Participants

Have participants talk about what they do and don't understand, what they do and don't like.

FIELD GUIDE 10.10: DIVERSIFYING LIVELIHOODS THROUGH MARKET LINKS

Diversifying livelihoods may be a good option for increasing resilience in the face of climate change challenges. There are many micro-lending and micro-enterprise programs in the developing world that are able to assist. However, not everyone is an entrepreneur. Another challenge is that frequently people think of a product they can make, begin making it, and have trouble finding customers. However, several simple things can be done to address these challenges.

Forming a Professional Association

The first one is to form an association of community members for the purpose of diversifying livelihoods and finding alternative forms of income generation. An association can give direction, consistency, management, and a single point of contact for a buyer. These services can be a benefit to association members who are not individually entrepreneurial. An association of producers can also provide buyers with larger quantities of product.

The first step in forming an association is to introduce the concept in a participatory workshop setting (as in Chapter 7). Frequently communities have groups of people pursuing similar activities. For example, there might be groups of weavers, artisans, or farmers. Ask your community contacts to approach one of these groups of community members about participating in a consciousness-raising workshop on alternative income generation.

In the workshop you can begin by introducing the rationale and importance of having a livelihoods-based association, and what benefits it could offer association members. Discuss how an initial step would be to set up an association management committee. Describe the types of skills that will need to be developed by committee members. Suggest appointing an interim committee who, over the course of six months, could develop the ultimate association structure, establish committee member responsibilities, ensure gender inclusiveness, set goals, and establish a plan. Have a discussion with the workshop participants to set a number of people that should serve on the committee—and then ask if there are participants who would be interested in sitting on the association management committee.

Surveying Businesses to Determine What They Need to Buy on a Routine Basis

An important activity for the committee will be to conduct a survey of businesses in the region that buy the kinds of products association members produce. Pick businesses that are purchasing products which are appropriate for the association's level of production capability. Ask them which products they most need. Ask if they would be interested in forming a market link with the association. Ask if they would be willing to partner in developing a training program so that the association members will be able to produce to their quality expectation.

An example could be an association of smallholder farmers who, through a survey, discovers a buyer who needs more squash for their wholesale markets in the city. Buyers such as these frequently need additional product, but do not have access to additional producers. It is inefficient for them to work with 20 individual smallholder farmers—they would prefer to work with a single contact person from an association of 20 producers. In order to develop a working relationship with an association, these buyers will often make an investment in farmer training and in micro-loans of seeds for planting. Agricultural produce buyers also frequently have extension agents who provide follow-up to the farmers.

There are different kinds of buyers that the association can look for. For example, there might be five or six buyers in an ascending market chain that stretches between smallholder producers and an exporter. The buyer immediately above the smallholder in the chain might be a large farmer who has market contacts: he could buy produce from the association to sell along with his own crop. Or, a buyer might be an intermediary, a food processor, a wholesaler, or the exporter themselves. Think positively: Successful sales organizations typically need additional produce to sell—and therefore need additional producers too.

Don't overlook other potential markets. Perhaps the open-air market in the next city has vendors which need products every week. Perhaps there are products that are in short supply that the association could sell themselves from a market stall. Before beginning production, however, make certain that there is a market. Your NGO might have board members with business experience who can help develop a market strategy or even make introductions to potential buyers.

Sales and Costs

Before committing to a relationship, you should have a clear definition of the product's value, its seasonality, the quantities that the buyer is intending to purchase, and the costs and time investment for production. Ask someone with bookkeeping skills to lay this out in a simplified presentation. This way, association members can determine how much money they could make through the sale of the product—and in what timeframe.

Establishing a Market Link and Setting up Vocational Training for New Products

Once you have identified an actual market for a product or service and solidified a relationship, members might need training in delivering that product. Solicit assistance from the buyer, the local government or perhaps an NGO for the training.

Association Management Committee Training

Get training for the association committee. They will need to know basic bookkeeping skills, basic management skills, and organizational skills. If they are the contact point for the buyer, they will also need to learn basic negotiation skills—and how to determine fair market value.

Workshop Lesson Plan for Diversifying Livelihoods through Market Links

- *Duration*: four hours (alternatively the workshop can be held in two-hour meetings).
- *Purpose*: Support participants in learning about the importance of diversifying livelihoods to strengthen the community:
  - Objective 1: Participants will understand diversifying livelihoods may be a good option for increasing resilience.
  - Objective 2: All will know that a professional association can help in diversifying livelihoods.
  - Objective 3: The association can conduct a survey of businesses as a method of developing sales.
  - Objective 4: Business partners can provide training, extension, and inputs.

Materials

- artist's drawings/posters; the scenes and people they contain should appear familiar to workshop participants;
- how-to cards without written words for workshop participants to take home;
- large sheets of newsprint and tape;
- colored markers.

Preparation

- Find a workshop location.
- 12 to 15 community members who do the same work (farmers, weavers, carpenters).

Activity 1: Introduction to the Benefits of Forming an Association. Duration 45 Minutes (Including a 15-Minute Ice Breaker)

Purpose

Participants will know a professional association can help in diversifying livelihoods.

What to Do

1 Ice breaker: Introductions. Sing a song or play a game.
2 Introduction to workshop: Tell the participants what they'll be able to do as a result of the lesson.
3 Discuss the challenges that community members face with low incomes.
4 Discuss the challenges that community members face with a lack of market links.
5 Discuss how diversifying livelihoods through market links can increase income
6 Discuss how an association can assist in organizing ways of linking to markets and diversifying livelihoods.

Workshop Participants

Have participants talk about what they do and don't understand, what they do and don't like.

Activity 2: Forming an Association. Duration 60 Minutes

Purpose

To understand how to form an association.

What to Do

1 discuss how an association can provide direction, consistency, and a single point of contact for a buyer;
2 discuss different ways that associations can be formed; they can begin informally and formalize over time;
3 discuss the skills that members of an association management committee will need;
4 suggest that an interim committee could over six months develop the ultimate association, set goals, and establish a plan;
5 ask the group to suggest a maximum number of management committee members;
6 ask for volunteers from the group to join an interim management committee.

Workshop Participants

Take 5 minutes to talk about what you do and don't understand, what you do and don't like.

Take a break for 15 minutes.

Introduction to Future Committee Activities

The next four activities are an overview and consciousness-raising introduction to activities that committee members can pursue over the next 12 months.

Activity 3: Surveying Businesses. Duration 20 Minutes

Purpose

To understand that an association can survey appropriate businesses in a search for business partners.

What to Do

1 develop a survey for businesses that buy products on a routine basis that association members make;
2 ask participants for ideas on businesses or markets that the association could approach with the survey;
3 to facilitate brainstorming, expand the definition of businesses to include neighborhood, local, regional, and national, markets. Think in terms of retail, wholesale, suppliers, manufacturers, processors, and exporters;
4 conduct a survey of businesses that buy products on a routine basis that association members make;

5 negotiate with the most appropriate businesses to form a market link;
6 ask if the business would be willing to make an investment in inputs and a producer training program.

Workshop Participants

Take 5 minutes to talk about what you do and don't understand, what you do and don't like.

Activity 4: Who Could Provide Vocational Training for Producing New Products? Duration 20 Minutes

Purpose

To understand what alternatives there are for training.

What to Do

1 ask the participants for ideas about what types of training they feel they might need to or would like to have;
2 government agencies can be approached for vocational training programs;
3 NGOs can be approached for vocational training programs;
4 business partners can be approached for vocational training;
5 suppliers can be approached for vocational training;
6 ask the participants for ideas of who they might be able to approach that could provide training.

Activity 5: Training for the Association Management Committee. Duration 20 Minutes

Purpose

To understand that the management committee will need training too.

What to Do

1 ask interim committee members what types of training they feel they might need to have to develop the committee;
2 they will need to know basic bookkeeping skills;
3 they will need to know basic management skills;
4 if they are the contact with the buyer, they will need to learn basic negotiation skills—and how to determine fair market value;
5 they will need training in transparency and in gender equity;
6 ask the participants for ideas of who they might be able to approach that could provide training.

Workshop Participants

Take 5 minutes to talk about what you do and don't understand, what you do and don't like.

Activity 6: How to Get Started? Duration 20 Minutes

Purpose

Participants will learn how to develop an initial plan for getting started on this program.

What to Do

First, schedule the next committee meeting where the committee will begin the following six-month organizational process:

1. *Organizational Development.* Interim committee members will:
   - a seek guidance for six months in developing an association and managing a committee;
   - b give responsibility for specific activities to specific committee members.
2. *Business Development.* The committee will develop an outline of the next steps to take:
   - a begin the development of a 12-month plan and a set of goals for business development;
   - b look at the strengths of the association members;
   - c identify appropriate businesses to approach;
   - d develop a survey form;
   - e conduct a trial survey with one business to see if the survey form works and to see if they learn ways to improve it;
   - f negotiate with the most appropriate businesses to form a market link;
   - g ask if the business would be willing to make an investment in inputs and a producer training program.

Activity 7: Conclusion. Duration 30 Minutes

Purpose

To reinforce what has been learned.

What to Do

1. discuss and review what has been learned;
2. reinforce the principles of forming an association as an aid to diversifying livelihoods and forming market links;
3. review the development of the management committee and of the association;
4. review the development of the initial 12-month plan for business development.

Workshop Participants

Take 10 minutes to talk about what you do and don't understand, what you do and don't like.

Chapter 10 Resources

Course Downloads

Go directly to this book's webpage, TimMagee.net/field-guide-to-cba/ to download and link to resources. On the book's webpage you can download the lesson plans that were provided for several of the book's field assignments and adapt them to your specific activity. You can also link to a number of practical handbooks and documents that were recommended in the field assignments. You can also see a compilation of these documents in the chapter called Recommended Resources.

On the book's webpage are several collections of specialized links to handbooks, manuals, and documents arranged by theme and by sector. Here you can find links to topics as diverse as:

- agriculture;
- soil and water conservation for agriculture;
- water harvesting for agriculture;
- household-level water harvesting;
- CBA techniques for agriculture;
- family gardens and nutrition;
- small island developing states;
- emergency preparedness and disaster risk reduction;
- participatory mapping;
- energy;
- livelihood diversification;
- health and hygiene;
- participatory techniques for working with communities.

Appendix

Background

People from 129 countries have used the techniques illustrated in this book to develop field projects having an impact on over 200,000 people.

The Evolution of the Book

The idea for providing training in cutting-edge development techniques for designing and implementing sustainable projects began as a series of training workshops that I held in Guatemala over several years. Participants actually worked within villages developing projects based upon participatory needs assessments they conducted with community members.

Two and a half years ago, the Center for Sustainable Development (CSDi) launched their online learning program using online courses I developed that faithfully followed the field component of these training workshops. People who participate in the online courses, and have access to communities, actually use the course materials in the field to develop real projects in real communities. Hence the moniker: online *field* courses.

This book, the *Field Guide to Community Based Adaptation* was developed from a CBA online diploma program that takes eight months to complete (four two-month courses). This is the right pace for busy professionals working in partnership with busy community members. By comparison, if you complete one chapter's field assignment per month, you will be working essentially at the same pace as the online students.

We were surprised when we saw how many people were enrolling in these courses. In the program's first 30 months—January 2010 until June 2012—people from 129 different countries and 380 organizations enrolled in 800 CSDi online courses to develop projects having an impact on over 200,000 people.

This has been an extraordinary learning experience for us—not only in being able to work with such a diversity of cultures—but in being able to work with over 200 different kinds of project activities. Consequently, this book has been able to take advantage of the rewarding experiences that we have had with these international students and their multi-contextual projects.

Here is a summary of the countries that the students have come from, the variety of project activities and solutions which they've worked with, and the names of the organizations they have worked for.

129 Countries

There are 129 countries where course participants live and work.

Afghanistan, Angola, Argentina, Australia, Austria, Azerbaijan, Bangladesh, Barbados, Belgium, Belize, Benin, Bolivia, Botswana, Brazil, British Virgin Islands, Bulgaria, Burkina Faso, Cambodia, Cameroon, Canada, Chad, Chile, China, Columbia, Costa Rica, Croatia, Czech Republic, Denmark, Djibouti, Dominica, Dominican Republic, Ecuador, Egypt, Ethiopia, Fiji, Finland, France, Georgia, Germany, Ghana, Grenada, Guatemala, Guyana, Haiti, Honduras, India, Indonesia, Ireland, Italy, Ivory Coast, Jamaica, Japan, Kazakhstan, Kenya, Kosovo, Kyrgyzstan, Lebanon, Lesotho, Liberia, Madagascar, Malawi, Malaysia, Mali, Mauritius, Mexico, Moldova, Monaco, Mongolia, Montserrat, Morocco, Mozambique, Myanmar, Namibia, Nepal, the Netherlands, New Caledonia, New Zealand, Nicaragua, Niger, Nigeria, Norway, Pakistan, Panama, Papua New Guinea, Paraguay, Peru, the Philippines, Poland, Portugal, Puerto Rico, Qatar, Reunion, Romania, Russia, Rwanda, Samoa, Senegal, Serbia, Sierra Leone, Singapore, Slovakia, Solomon Islands, Somalia, Somaliland, South Africa, South Sudan, Spain, Sri Lanka, Sudan, Swaziland, Switzerland, Tajikistan, Tanzania, Thailand, Timor Leste, Trinidad and Tobago, Tristan da Cunha, Turkey, Uganda, Ukraine, United Arab Emirates, the United Kingdom, the United States, Uruguay, Venezuela, Vietnam, Yemen, Zambia, Zimbabwe.

Students' Project Activities

Students' project activities focus on over 215 different types of challenges that communities face.

Activities for the Elderly
Acutely Malnourished Children
Adaptation to Climate Change
Adult Literacy
Advocacy
Advocacy for the Elderly
Afforestation
Agricultural Adaption
Agricultural Coping Strategies
Agricultural Income Generation
Agricultural Infrastructure
Agricultural Vulnerabilities
Agriculture
Agrobiodiversity
Agroecology
Agroecotourism
Agroforestry
Alternative Livelihoods
Alternative Transportation
Animal Husbandry
Aquaculture/Mariculture
Artificial Reefs
Bicycle Paths
Bio-engineering for Slope Stabilization
Capacity Building for Farmers
Capacity Building for Women
Cassava: Disease Resistant
Civic Leadership
Climate Smart Agriculture
Climate Smart Agroforestry
Coastal Management
Coffee Production
Cold Storage Fish Facility
Community Based DRR
Community Based Flood Adaptation
Community Based Flood Mitigation
Community Based Forestry
Community Development Committees
Community Health
Community Health Center
Community Health Promoters
Community Management Committees
Community Organization
Community Youth Activities
Composting
Composting Toilets
Conservation
Conservation Tillage
Coral Reef Preservation
Crop Diversification
Deaf in Zambia
Deforestation

Degraded Watersheds Restoration
Direct Marketing Agricultural Products
Disaster Preparedness
Disaster-Resistant School Construction
Disaster Risk Management
Drought Mitigation
Drug Abuse
Ecotourism
Education
Educational Advocacy
Elderly Care Facilities
Emergency Preparedness
Employment Services
Empowering the Elderly
Empowerment of Women
Entrepreneurialism
Environmental Sanitation
Environmental Services
Environmental Sustainability
Family Gardens and Nutrition
Family Planning
Farmer Association
Farmer Soil and Water Conservation
Farmer to Farmer Schools
Farmer Training
Farmer Water Conservation
Financial Literacy
Fisheries Management
Fishing Stock Stewardship
Flood Early Warning
Flood Mitigation
Flood Resilience Program
Flood Response
Food and Nutrition
Food Preservation
Food Processing
Food Security
Forest Conservation
Forest Restoration
Forest Stewardship
Fundraising
Garden Fence Installation
Greenhouse Agriculture
Habitat Conservation
Habitat Restoration
Hazard Mapping
Health
Health and Hygiene
Health and Traditional Knowledge
Home Gardens
Household Energy Efficiency
Human Rights
Hygiene and Sanitation Program
Improved Agricultural Practices
Improved Cook Stoves
Income Generation
Indoor Air Pollution
Irrigation
Land Degradation
Land Use Management
Latrines
Leadership
Linking Artisans to Markets
Linking Farmers to Markets
Livestock Association
Livestock Protection/Weather Extremes
Mangrove Restoration and Conservation
Marine Conservation Programs
Marine Resource Management
Market Development
Market Linkages
Market Survey
Marketing and Networking Skills
Marketing of Crops
Marketing of Livestock
Micro-Credit
Micro-Enterprise
Micro-Finance
Micro-Hydro
Micro-Irrigation
Micro-Savings and Loan Programs
Migration/Family Disintegration
Money Management
Monitoring and Evaluation
Mulching
Natural Resource Degradation
Natural Resource Management
Neighborhood Clean-up
Non-Timber Forest Products
Nutrition
Nutrition and Food Diversity
Organizational Strengthening
Orphans
Parental Involvement
Parental Responsibility

Participatory Afforestation
Participatory Forest Management Committees
Participatory Hazard Mapping
Participatory Mapping
Participatory Mapping of Water Resources
Participatory Monitoring and Evaluation
Participatory Reforestation
Participatory Tree Planting
Pastoral Land Use Management
People with Disabilities
Permaculture Center
Piped in Water/Wells/Infrastructure
Post-Harvest Management
Preserving Traditional Culture
Preserving Traditional Knowledge
Project Management
Projects by Category
Protection of Cultural Patrimony
Provide School Materials
Public Health
Rainforestation Farming
Rainwater Harvesting
Range Management
Recycling
REDD+
Reforestation
Refuse Management
Reproductive Health
Resilience to Extreme Weather Conditions
Resilience to Weather Variability
Restoration
Rice Farming
River Conservation
Safe Water
Sanitation
Savings Programs
School Drop Outs
School DRR
Sedimentation Reduction Programs
Sexual and Reproductive Rights
Small Animal Production
Social Assistance
Soil Conservation
Soil Restoration
Soil, Water, and Natural Resource Conservation
Solar Ovens
Solid Waste
Spring Protection
Storm Water Management
Supply Chains
Surface Water Management
Surveillance for Sustainable Resource Management
Sustainable Livelihoods
Sustainable Tourism
Synergistic Agroforestry
Tree Seedling Nurseries
Value Chains
Vocational Skills Training
Vocational Training
Water Conservation
Water Harvesting
Water Hygiene and Purification
Water Pump Installation and Maintenance
Water Purification
Water Supply Programs
Water Use Management
Youth Programs and Mentoring

Students have worked with 380 organizations

Academy for Educational Development
Act for Peace
Action Center for City Development
Action Contre la Faim
Action Internationale de Développement Intégral
Action on Disability and Development
Adventist Development & Relief Agency
Advocates & Trainers for Women's Welfare
Advancement & Rights
Aflatoun
African Development Bank
African Development by African Diaspora
Agencia Suiza para el Desarrollo
Albertine Rift Conservation Society

Andrews University
Asia Foundation China
Asia Network for Sustainable Agriculture & Bioresources
Asian Development Bank
Asian Institute of Technology
Asociación Nuestros Ahijados
Asociación BPD Guatemala
Asociación El Shaddai
Asociación para la Naturaleza y el Desarrollo Sostenible
Asociación Rijatzul Q'ij – Semilla de Sol
Association 4D
Association for Sustainable Community Development
Association Intercooperation Madagascar
Association Lead Tchad
Association of Private Water Operators Uganda
Association Rif for Solidarity and Dev.
B9 Plastics
Atlantida MF
Australian Agency for International Development
Australian Dept of Climate Change and Energy Efficiency
Bangor University
Benevolent Institute of Development Initiatives
Beyond Poverty
Bosque Modelo Atlantida
Bosque Modelo de las Tierras Adjuntas
Bosque Modelo Jujuy
Bosque Modelo Urbión
Bosques Amazonicos
CAB International
Caleb Management Services Ltd
California Dept of Public Health
Camco Global
Campo-Ma'an Model Forest
Canadian International Development Agency
CARE International
CARE Kenya
CARE Nepal
Caribbean Community Climate Change Centre
Caribbean Community Secretariat
Caribbean Sustainable Institute
Catholic Arch Diocesan Rural/Urban Development Program
Cedre Consult
Center for Alternative Resources
Center for Environmental Consulting & Eng.
Center for International Forestry Research
Center for Nursery Development
Center for Security and Peace Studies, Gadjah Mada University
Center for Sustainable Development
Centre for People's Forestry
Centre for Rural Technology, Nepal
Centre for Sustainable Rural Development Vietnam
Cerrejon Foundation for Water
CGIAR
Champions for Change
Change for Children Association
Childfund International
China Agricultural University
Christian Mission Aid
Christian Reformed World Relief Committee
CIFOR
Clean Up the World
Climate Concern
Clinton Foundation
Coastal Oceans Research and Development in the Indian Ocean East Africa
Community Capacity Building Initiative
Community Disaster Management
Community Enterprise Solutions
Community Research and Development Centre
CONAP Consejo Nacional de Áreas Protegidas
Concern Universal
Concern Universal, Bangladesh
Concern University, Bangladesh
Concern Worldwide
CONICET Argentina
Conservation & Fisheries BVI Gov.

Conservation International
Consultative & Research Center for Natural Resource Management
Consumers International
Copperbelt Indigenous Peoples Land Rights Network
Corporación Aldea Global
Corporación Autónoma Regional Cauca
Corporación Nacional Forestal (CONAF)
Corporación para la Investigación
Costa Rica Entity of Accreditation
Counterpart Cameroon
Counterpart International
Cranbery Consulting Ltd.
Cross River State Forestry Commission
CRWRC
Dalhousie University
David and Lucile Packard Foundation
Deakin University
Delft University of Technology
Delta Women
Delta Women Nigeria
Department of Fisheries Zanzibar
Development Alternative Inc.
Development Effort for Self Help
Development Plus Uganda
Dezman Investments
Divya Jyoti Jagrati Sansthan
Dominica National Fairtrade Organization
Eastern Ontario Model Forest
Echo Group
eco2ro
Ecocontact
Ecoisle Consulting
EcoLogic Development Fund
Ecological Living In Action
ECOTRAC Consulting
Education Development Center, Inc. Philippines
Ekuri Initiative
El Observatorio Venezolano de Derechos Humanos de las Mujeres
ENVIRO-DEV
EnvironHealth Systems Limited
Environment and Public Org.
Environmental Resources Management
Escuela Organización Industrial
Etnia Verde
European Union
European Volunteer Service
Evangelischer Entwicklungsdienst
Evergreen Agriculture Tanzania
Fairtrade International
Family Health International
FAO
FARM-Africa
FEM International
FLACSO
Fletcher School of Law and Diplomac
Fondo de Atención y Prevención de Emergencias
Ford Foundation, East Africa Region
Forestry Commission Community
Forestry Unit
Frankfurt School of Finance & Management
Freshwater Action Network South America
Fundación Gaia Pacha
Fundación Prosierra Nevada
Fundación Tradiciones Mayas
Genivar
Georg Kraus Stiftung
GERES
German International Cooperation
German Red Cross
German Society for Development (GIZ)
Global Gender and Climate Alliance (GGCA)
GMS Environment Operations Center
Government of British Columbia
Green Belt Movement International Europe
Green Farms Initiative
Greenlight Biofuels Indonesia
Grupo FARO
Guyana Mangrove Restoration Project
Haitian American Liaison
Hanalei Watershed Hui
Handicap International
Hatfield Consultants

Hayata Destek Association
Hivos
Humanity Exchange
Humboldt Universität zu Berlin
IBS Kolkata
ICRAF
IDEA Int. U. Inca Garcilazo
IDS University of Sussex
IFMR-Centre for Development Finance
IFRC Myanmar
Imperial College London
Independent Consultant
Initiatives for Development
Institute of Development Studies
Instituto de Investigación Universidad y Region IIUR
Instituto Ipanema
Intensify Global Foundation
Inter-American Development Bank
Intercooperation Switzerland
International Center for Integrated Mountain Development
International Conflict and Security Consulting Ltd
International Finance Corporation
International HIV/AIDS Alliance, South Sudan
International Labour Organization
International Model Forest Network
International Organization for Migration
International Renewable Resource Institute
Iringa Development of Youth, Disabled & Children
IT Power Group
IUCN
Jerusalem Children and Community Development Organization
Joint Aid Management International
Joseph Fourier University, Grenoble, France
Jubbalandese Charity Centre
KPMG
Kuruka Maisha Foundation
Kyoto Club
La Paz on Foot
La Plaza de Cultura y Artes
Lake Albert Agricultural and Forestry Imitative
Lake Simcoe Region Conservation Authority
Lambeth Social Services
Lay Volunteers International Association Ethiopia
Leadership for Environment and Development
Legal Earth International
Ligingi Community Learning Centre
Livelihoods and Forestry Programme Nepal
Lok Sanjh Foundation
Madagascar National Parks Project COFAM
Mancomunidad de Municipios del Centro de Atlantida
Mauritius Meteorological Services
mc2group Management Coaching & Consulting Group
McGill University
Médecins du Monde
Medical Care Development International
Métis National Council
Migratory Wildlife Network
Ministry of Agriculture and Cooperatives
Ministry of Agriculture and Natural Resource Nigeria
Ministry of Environment Georgia
Ministry of Environment Morocco
Ministry of Health Kenya
Ministry of Justice UK
Mission Guatemala
MIT Department of Urban Studies and Planning
Moldova State University
Mongolian Ministry of Environment
Montserrat National Trust
Mulanje Renewable Energy Agency
Nascent Solutions Inc.
National Commission for Refugees
National Disaster Management Agency
National Institute for Spatial Research, (INPE)
National Research Council

National University of Costa Rica
National Water and Sewerage Corporation
Nature Conservancy
NEXUS
NGO Promo Antananarivo Madagascar
North-South Environmental Inc.
Offsetters
Olof Palme Peace Foundation
One Justice Project
Organización Mundial de Conservación
Organization for Environment and Sustainable Development
Oxfam GB
Oxfam GB Uganda
Oxfam Quebec
Oxford University
Pacific Australia Climate Change Science and Adaptation Planning Program
Panos Institute Southern Africa
Panos London
Peruvian National Service of Meteorology and Hydrology
Petrobras Research Center
Plan Canada
Plan-International, South Sudan
Platform for Agrobiodiversity Research
PNUD Perú
PNUD Burkina Faso
Pontificia Universidad Católica del Ecuador
PREVAL
Prochorus Africa Services Uganda
Progressio Development Worker
Public Health Institute
Qatar Foundation
Rainforest Alliance
Regional Centre for Development Cooperation
Remote Area Medical
Research Institute for Humanity & Nature
Revelation Life
Rights and Resources Group
RIMS Nepal
Risaralda MF
ROAR Global Media
Rural Family Development Association
Save the Children Ethiopia
Save the Children India
Save the Children UK
Save the Children US
Sécretariat Permanent du Conseil National pour l'Environnement et le Développement Durable Burkina Faso
Self-employed Consultant
SGS do Brazil Ltda
Sierra Club Quebec
Sistren Theatre Collective, Jamaica
Sivas-Erzincan Development Project
Sustainable Arts and Culture Foundation
Sustainable Health Enterprises
Susu Mamas PNG
Swedish Cooperative Centre for Eastern Africa
Tecnológico de Monterrey
Telitech Computers
The Nature Conservancy
TRAFFIC Malaysia
Triangle Génération Humanitaire
TRINETRA Youth Vision
Tristan Conservation Department
U. Sergio Arboleda, Universidad Externado de Colombia
Uganda Coalition for Sustainable Development
Uganda Martyrs University-Nkozi
Uganda Police Force
Uganda Wildlife Authority
UN FAO
UN Mission in Liberia
UN Office on Drugs and Crime
UN Women
UN World Food Programme Rome
UN World Tourism Organization
UNAMID
UNDP
UNDP Bangkok
UNDP Barbados
UNDP EEG
UNDP Honduras Early Recovery Project
UNDP Solomon Islands
UNEP

UNEP Risoe Centre DTU
UNESCO
UNESCO Ministerio del Desarrollo Social y Combate al Hambre Brasil
UNICEF
UNICEF Nepal
UNICEF Peshawar
Union Aid Abroad, APHEDA
United Nations African Union Mission in Darfur (UNAMID)
United Nations Development Programme
United Nations International Strategy on Disaster Reduction
United Nations Volunteers
Universidad Autonoma de Barcelona
Universidad de Buenos Aires
Universidad de Salamanca
Universidad Kassel
Universidad Michoacana
Universidad Nacional Autónoma de Mexico
Universidad Nacional, Heredia Costa Rica
Universidad Tecnológica de Pereira
Université Libre de Kigali
University of Cambridge
University of Delaware
University of Edinburgh
University of Florida
University of Nigeria Nsukka
University of Ottawa
University of Sussex
University of the Philippines Baguio
University of the Witwatersrand S. Africa
University of Windsor
University of Yaounde
University of Zambia Lusaka
UNOCHA Ethiopia
UNODC
Urban Development Corporation
Urban Gardens Program Ethiopia
US Environmental Protection Agency
US Fulbright Program
Utrecht University
Véterinaires Sans Frontières Belgium
Virgin Blue Airlines
Virtuous Values Network Africa
Vumilia CBO
Wamumbi Orphan Care
Winrock International
WITS University
World Agroforestry Center
World Bank
WWF
WWF European Policy Office
WWF UK
Yale University
Youth Link Somalia

Recommended Resources

People from 170 countries subscribe to the CSDi newsletter. CSDi's information-rich newsletter contains up-to-the-minute reports on student field projects. You can see the complete collection of CSDi newsletters at: http://www.csd-i.org/newsletters/

You can subscribe to CSDi's newsletter at: http://www.csd-i.org/subscribe/

Learn about the full range of CSDi's online field courses at: http://www.csd-i.org/

Recommended Resources

Website addresses change frequently. Simply enter this book's webpage for current links to resources, or enter the author's name, the organization's name and the document's name into your web browser to find the most current link.

Bonita, M. and Payuan, E. (2001) *Village Forestry Handbook*, Vientiane: Ministry of Agriculture and Forestry Lao People's Democratic Republic. Online. Available at: http://www.mekonginfo.org/assets/midocs/0003105-environment-village-forestry-handbook.pdf (accessed 21 April 2012).

CARE International (2009) *Climate Vulnerability and Capacity Analysis Handbook*. Atlanta, GA: Cooperative for Assistance and Relief Everywhere, Inc. Online. Available at: http://www.careclimatechange.org/files/adaptation/CARE_CVCAHandbook.pdf (accessed 12 April 2012).

CARE International (2010) *Community-Based Adaptation Toolkit*, Atlanta, GA: Cooperative for Assistance and Relief Everywhere, Inc. Online. Available at: http://www.careclimatechange.org/files/toolkit/CARE_CBA_Toolkit.pdf (accessed 12 April 2012).

CARE International (undated) *Framework of Milestones and Indicators for Community Based Adaptation*. Atlanta, GA: Cooperative for Assistance and Relief Everywhere, Inc. Online. Available at: http://www.careclimatechange.org/files/toolkit/CBA_Framework.pdf (accessed 12 April 2012).

Carloni, S. (2005) *Rapid Guide for Missions: Analyzing Local Institutions and Livelihoods*, Rome: Food and Agriculture Organization of the United Nations. Online. Available at: ftp://ftp.fao.org/docrep/fao/008/a0273e/a0273e00.pdf (accessed 28 May 2012).

Cash, D., Clark, W., Alcock, F., Dixon, N., Eckley, N., Guston, D. Jager, J., and Mitchell, B. (2003) *Knowledge Systems for Sustainable Development*, Washington, DC: Proceedings of the National Academy of Sciences of the United States of America, PNAS, Volume 100, No. 100, 8 July 2003. Online. Available at: http://www.pnas.org/content/100/14/8086.full.pdf+html?sid=7c343bfa-582c-425e-ac8a-84739494c330 (accessed 12 April 2012).

Chatty, D., Baas, S., and Fleig, A. (2003) *Participatory Processes towards Co-Management of Natural Resources in Pastoral Areas of the Middle East. Module II: Introducing Participatory Approaches Methods and Tools*, Rome: Food and Agriculture Organization of the United Nations. Online. Available at: ftp://ftp.fao.org/docrep/fao/006/ad424e/ad424e00.pdf (accessed 28 May 2012).

Dayal, R., van Wijk, C., and Mukherjee, N. (2000) *Methodology for Participatory Assessments with Communities, Institutions and Policy Makers*, Washington, DC: World Bank. Online. Available at: http://bscw.ihe.nl/pub/nj_bscw.cgi/d2220629/DayalMethodologyforParticipatoryAssessments.pdf (accessed 28 May 2012).

FARM-Africa (2007) *The Key Steps in Establishing Participatory Forest Management: A Field Manual to Guide Practitioners in Ethiopia*, Addis Ababa: FARM-Africa Ethiopia. Online. Available at: http://www.farmafrica.org.uk/resources/Key%20Steps%20in%20Establishing%20Participatory%20Forest%20Management2.pdf (accessed 21 April 2012).

IFAD (2009) *Good Practices in Participatory Mapping*, Rome: International Fund for Agricultural Development. Online. Available at: http://www.ifad.org/pub/map/PM_web.pdf (accessed 12 April 2012).

IIED (2009) *Mapping for Change: Practice, Technologies and Communication*, Participatory Learning and Action 54, London: International Institute for Environment and Development. Online. Available at: http://www.planotes.org/pla_backissues/54.html (accessed 12 April 2012).

Jain, S. and Polman, W. (2003) *Training Module on Participatory Community Monitoring and Evaluation*, Rome: Food and Agriculture Organization of the United Nations. Online. Available at: http://www.fao.org/docrep/006/AD346E/ad346e0e.htm (accessed 21 April 2012).

Jayakaran, R. (2002) *Ten Seed Technique*, China: World Vision International. Online. Available at: http://ravijayakaran.com/books.htm (accessed 28 May 2012).

Kato, E., Ringler, C., Yesuf, M., and Bryan, E. (2009) *Are Soil and Water Conservation Technologies a Buffer Against Production Risk in the Face of Climate Change? Insights from Ethiopia*, Washington, DC: International Food Policy Research Institute, IFPRI Discussion Paper No. 871. Online. Available at: http://www.ifpri.org/sites/default/files/publications/rb15_17.pdf (accessed 21 April 2012).

Krantz, L. (2001) *The Sustainable Livelihood Approach to Poverty Reduction*, Stockholm: Swedish International Development Cooperation Agency. Online. Available at: http://www.sida.se/Svenska/Om-oss/Publikationsdatabas/Publikationer/2003/september/The-Sustainable-Livelihood-Approach-to-Poverty-Reduction/ (accessed 28 May 2012).

Mathie, A. and Foster, M. (2006) *Participatory Monitoring and Evaluation: A Manual for Village Organizers*, Antigonish, NS, Canada: Coady International Institute, St. Francis Xavier University. Online. Available at: http://coady.stfx.ca/resources/abcd/SEWA%20PME%20Manual.pdf (accessed 21 April 2012).

Mitchell, T. and Tanner, T. (2006) *Adapting to Climate Change: Challenges and Opportunities for the Development Community*, Teddington: Tearfund. Online. Available at: http://www.tearfund.org/webdocs/website/Campaigning/policy%20and%20research/Adapting%20to%20climate%20change%20discussion%20paper.pdf (accessed 12 April 2012).

NGO Programme Karnataka-Tamil Nadu (2005) *Participatory Monitoring and Evaluation: Field Experiences*, NGO Programme Karnataka-Tamil Nadu Series 1 Intercooperation Delegation, Hyderabad, India. Online. Available at: http://www.sswm.info/sites/default/files/reference_attachments/Intercooperation%202005%20Participatory%20Monitoring%20And%20Evaluation.pdf (accessed 12 April 2012).

Regmi, B., Morcrette, A., Paudyal, A., Bastakoti, R., and Pradhan, S. (2010) *Participatory Tools and Techniques for Assessing Climate Change Impacts and Exploring Adaptation Options*, Kathmandu: Livelihoods and Forestry Programme. Online. Available at: http://www.lfp.org.np/publications.php?id=34 (accessed 28 May 2012).

Stalker, L. (2001) *Why Some Village Water and Sanitation Committees Are Better than Others*, New Delhi: World Bank Water and Sanitation Program-South Asia. Online. Available at: https://www.wsp.org/wsp/sites/wsp.org/files/publications/327200744509_saybetter.pdf (accessed 21 April 2012).

Theis, J. and Grady, H. (1991) *Participatory Rapid Appraisal for Community Development*, London: International Institute for Environment and Development. Online. Available at: http://pubs.iied.org/pdfs/8282IIED.pdf (accessed 12 April 2012).

UNDP, Bureau of Development Policy (2010) *Designing Climate Change Adaptation Initiatives: A UNDP Toolkit for Practitioners*, New York: United Nations Development Programme. Bureau of Development Policy. Environment and Energy Group. Online. Available at: http://www.undp.org/content/undp/en/home/librarypage/environment-energy/low_emission_climateresilientdevelopment/designing-adaptation-initiatives-toolkit.html (accessed 12 April 2012).

United Nations Development Programme (2010) *Gender, Climate Change and Community-Based Adaptation: A Guidebook for Designing and Implementing Gender-Sensitive Community-Based Adaptation Programmes and Projects*. Online. Available at:http://www.undp-adaptation.org/projects/websites/docs/KM/PublicationsResMaterials/Gender_Climate_Change_and_Community_Based_Adaptation_%282%29.pdf (accessed 12 April 2012).

V&A Programme (2009) Vulnerability and Adaptation Experiences from Rajasthan and Andhra Pradesh: Community Based Institutions, New Delhi: V&A Program. Online. Available at: http://www.intercooperation.org.in/images/Climate%20Change%20-%20Case%20study%20on%20Community%20based%20Institutions.pdf (accessed 21 April 2012).

World Bank (n.d.) World Bank Climate Change Portal. Washington, DC: World Bank. Online: Available at: http://sdwebx.worldbank.org/climateportal/index.cfm (accessed 12 April 2012).

Bibliography

Adaptation Fund (2010) *Proposal for Guatemala*. Washington, DC: The Adaptation Fund Board. Online. Available at: http://adaptation-fund.org/system/files/AFB.PPRC_.2.6%20Proposal%20for%20Guatemala.pdf (accessed 12 April 2012).

Bhattacharjee, L., Phithayaphone, S., and Nandi, B. K. (2006) *Home Gardens Key to Improved Nutritional Well-Being*, Bangkok: Food and Agriculture Organization of the United Nations, Regional Office for Asia and the Pacific. Online. Available at: ftp://ftp.fao.org/docrep/fao/meeting/011/ag101e/ag101e00.pdf (accessed 21 April 2012).

Burgess, A. and Glasauer, P. (2004) *Family Nutrition Guide*, Rome: Food and Agriculture Organization of the United Nations. Online. Available at: http://www.fao.org/docrep/007/y5740e/y5740e00.HTM (accessed 21 April 2012).

English, R. M., Badcock, J. C., Giay, T., Ngu, T., Waters, A-M., and Bennett, S. A. (1997) Effect of nutrition improvement project on morbidity from infectious diseases in preschool children in Vietnam: comparison with control commune, *BMJ* 315, 1 November 1997.

FAO (2001) *Improving Nutrition through Home Gardening: A Training Package for Preparing Field Workers in Africa*, Rome: Food and Agriculture Organization of the United Nations. Online. Available at: http://www.fao.org/docrep/003/x3996e/x3996e00.htm (accessed 21 April 2012).

FAO (2010) *"Climate-Smart" Agriculture*, Rome: Food and Agriculture Organization of the United Nations. Online. Available at: http://www.fao.org/fileadmin/user_upload/newsroom/docs/the-hague-conference-fao-paper.pdf (accessed 21 April 2012).

Iannotti, L., Cunningham, K., and Ruel, M. (2009) *Improving Diet Quality and Micronutrient Nutrition: Homestead Food Production in Bangladesh*, Washington, DC: International Food Policy Research Institute (IFPRI). Online. Available at: http://www.ifpri.org/sites/default/files/publications/ifpridp00928.pdf (accessed 21 April 2012).

Kato, E., Ringler, C., Yesuf, M., and Bryan, E. (2009) *Are Soil and Water Conservation Technologies a Buffer Against Production Risk in the Face of Climate Change? Insights from Ethiopia*, Washington, DC: International Food Policy Research Institute, IFPRI Discussion Paper No. 871. Online. Available at: http://www.ifpri.org/sites/default/files/publications/rb15_17.pdf (accessed 21 April 2012).

Kropp, J. and Scholz, M. (2009) *Climate Change Information for Effective Adaptation: A Practitioner's Manual*, Eschborn: Deutsche Gesellschaft für Technische Zusammenarbeit (GTZ). Online. Available at: http://www2.gtz.de/dokumente/bib-2009/gtz2009-0175en-climate-change-information.pdf (accessed 12 April 2012).

Lennox, J. (ed.) (2010) *The Economics of Climate Change in Central America: Summary 2010*. New York: United Nations Economic Commission for Latin America and the Caribbean (ECLAC). Online. Available at: http://www.eclac.org/cgi-bin/getProd.asp?xml=/publicaciones/xml/9/41809/P41809.xml&xsl=/mexico/tpl-i/p9f.xsl&base=/mexico/tpl/top-bottom.xslt (accessed 12 April 2012).

Reid, H. Alam, M., Berger, R. Cannon, T., Huq, S., and Milligan, A. (2009) *Community-Based Adaptation to Climate Change: An Overview*. Participatory Learning and Action 60, London:

International Institute for Environment and Development. Online. Available at: http://www.iied.org/pubs/pdfs/14573IIED.pdf (accessed 12 April 2012).

UNDP-WHO (1994) *Food, Water and Family Health: A Manual for Community Educators*, Geneva: World Health Organization. Online. Available at: http://www.who.int/water_sanitation_health/hygiene/settings/wsh9204.pdf (accessed 21 April 2012).

United Nations General Assembly (1987) *Report of World Commission on Environment and Development, Our Common Future*. Online. Available at: http://www.un-documents.net/a42r187.htm (accessed 28 May 2012).

Waddington, H., Snilstveit, B., White, H., and Fewtrell, L. (2009) *Water, Sanitation and Hygiene Interventions to Combat Childhood Diarrhoea in Developing Countries*, New Delhi: International Initiative for Impact Evaluation, Synthetic Review 001. Online. Available at: http://www.3ieimpact.org/admin/pdfs2/17.pdf (accessed 21 April 2012).

Wilson-Grau, R. (2008) Customising definitions of outputs, outcomes and impact, unpublished. Online. Available at: http://www.outcomemapping.ca/resource/resource.php?id=189 (accessed 23 April 2012).

World Bank (2009) *Guatemala Country Note: Climate Change Aspects in Agriculture*. Washington, DC: World Bank. Online: Available at: http://siteresources.worldbank.org/INTLAC/Resources/Climate_GuatemalaWeb.pdf (accessed 12 April 2012).

Zwane, A. P. and Kremer, M. (2007) What works in fighting diarrheal diseases in developing countries? A critical review, *World Bank Research Observer* 22(1): 1–24.

Index

Figures are indicated by italic page numbers.